Holt
Mathematics

Chapter 10 Resource Book

HOLT, RINEHART AND WINSTON

A Harcourt Education Company

Orlando • **Austin** • New York • San Diego • London

ISBN 0-03-078307-0

1 2 3 4 5   170   09 08 07 06

CONTENTS

Holt Mathematics

Holt Mathematics

Date ______________

Dear Family,

In this chapter, your child will learn how to identify and work with three-dimensional figures, including finding the volume and surface area of cylinders, pyramids, and other three-dimensional figures. Understanding these concepts can help us appreciate famous structures such as the pyramids of ancient Egypt, and the study of volume and surface area is important in architecture and computer-aided design.

Your child will learn to identify various three-dimensional figures such as a **polyhedron,** a figure whose surfaces are polygons.

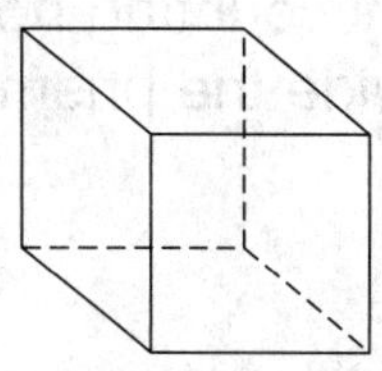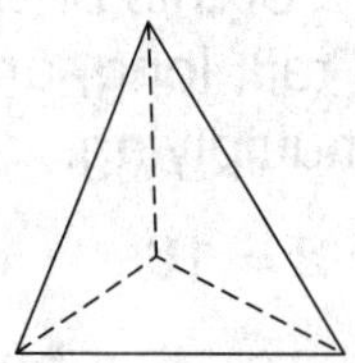

A **prism** is a polyhedron that has two parallel congruent bases and remaining faces that are parallelograms.

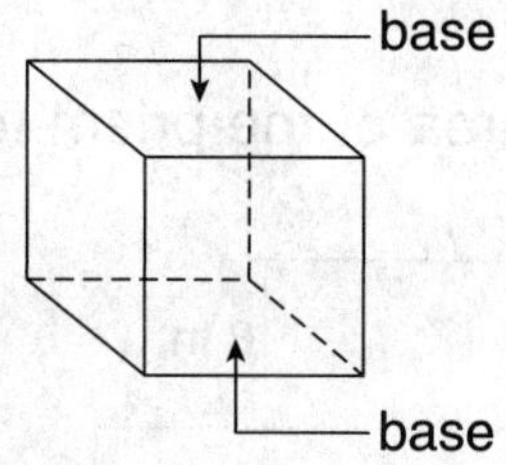

A **pyramid** is a polyhedron that has a vertex and a base at opposite ends, with remaining faces that are triangles.

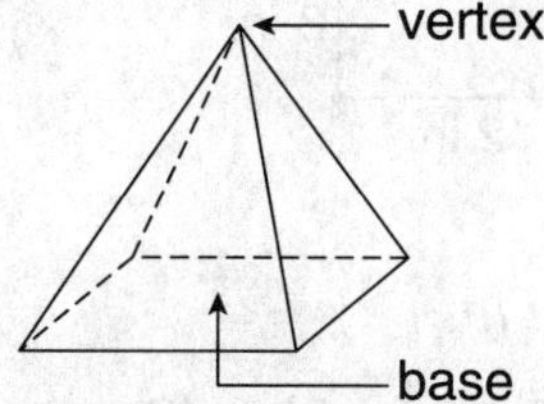

A **cylinder** has two congruent circles as bases connected by a curved surface.

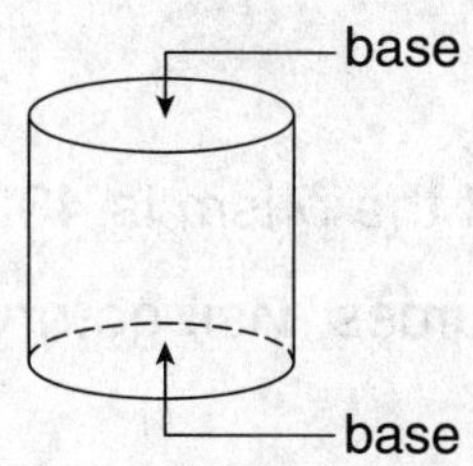

A **cone** has only one circular base and a curved surface that comes to a point called the vertex.

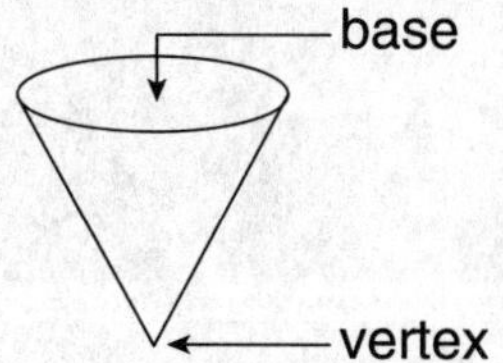

Holt Mathematics

Your child will learn to find the **volume of prisms and cylinders.** A cube that measures one centimeter on each side represents **1 cubic centimeter of volume.**

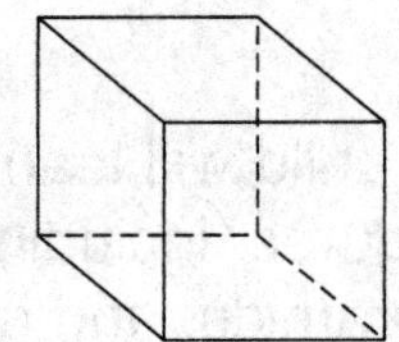

You can use cubes to find the **volume of a rectangular prism.**

Find how many cubes the prism holds and tell the volume of the prism. You can find the volume of this prism by counting how many cubes tall, long, and wide the prism is and then multiplying.

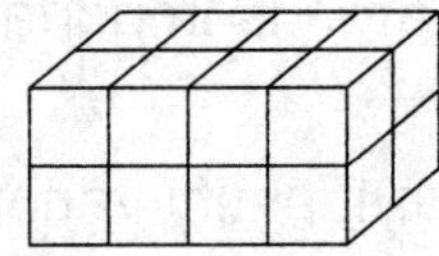

$4 \cdot 2 \cdot 2 = 16$

There are 16 cubes in the prism, so the volume is 16 cubic units.

Your child will learn to find the **surface area** of prisms, cylinders, and spheres.

Find the surface area of the prism formed by the net.

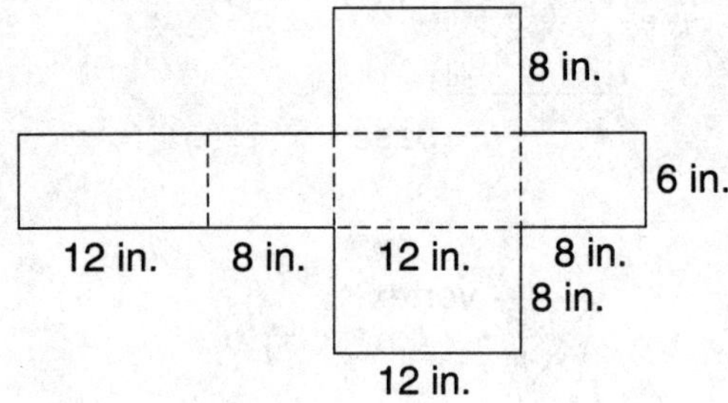

$S = 2\ell w + 2\ell w + 2\ell w$	*Use the formula.*
$S = (2 \cdot 12 \cdot 8) + (2 \cdot 12 \cdot 6) + (2 \cdot 8 \cdot 6)$	*Substitute.*
$S = 192 + 144 + 96$	*Simplify.*
$S = 432$	

The surface area of the prism is 432 square inches.

For additional resources, visit go.hrw.com and enter the keyword MS7 Parent.

Holt Mathematics

Practice A
Introduction to Three-Dimensional Figures

**Identify the base of each prism or pyramid. Then choose the
name of the prism or pyramid from the box.**

rectangular prism	square pyramid	triangular prism	pentagonal prism
square prism	triangular pyramid	hexagonal prism	rectangular pyramid
hexagonal pyramid	pentagonal pyramid		octagonal prism

1. 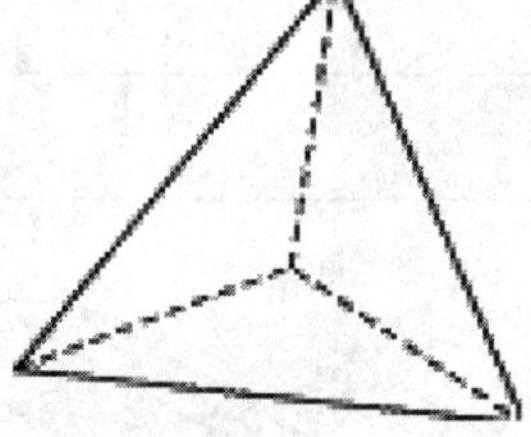

____

2. 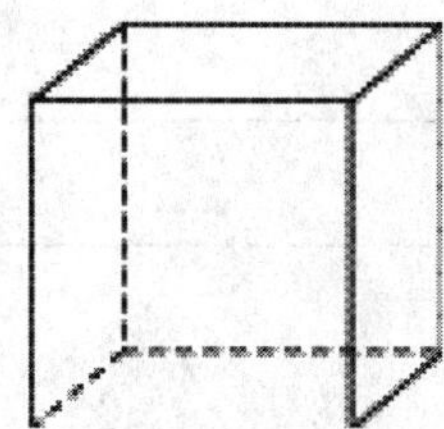

____

3. 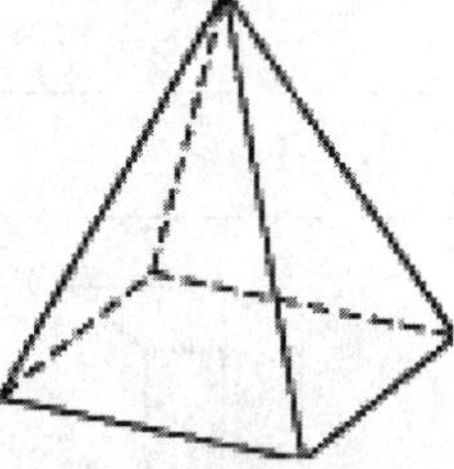

____

4. 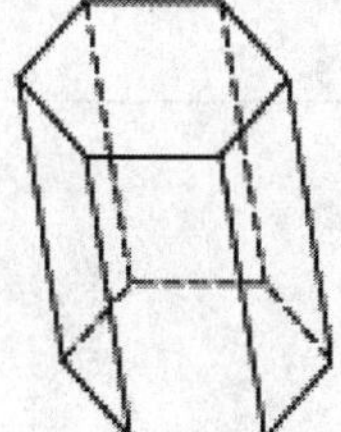

____

5. 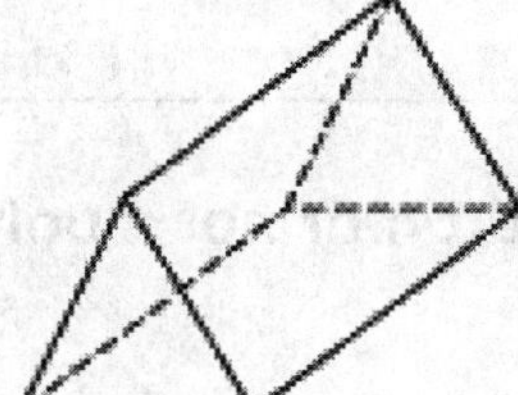

____

6. 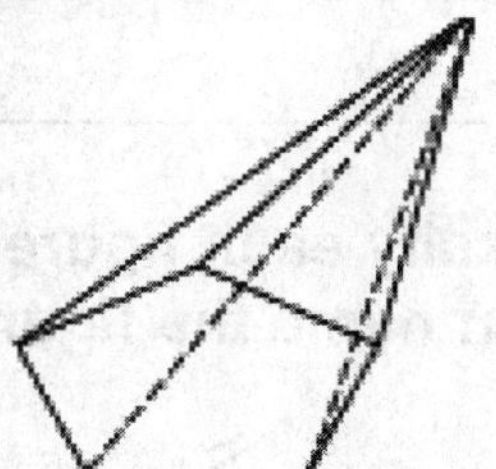

____

**Classify each figure as a polyhedron or not a polyhedron.
Then name the figure.**

7. 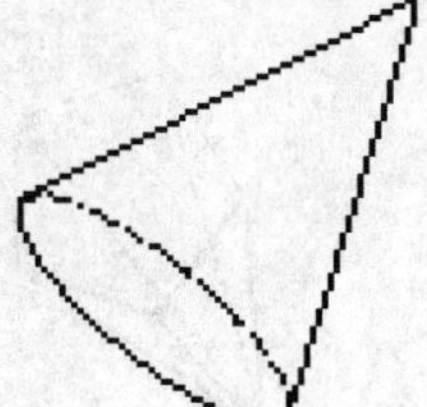

____

8. 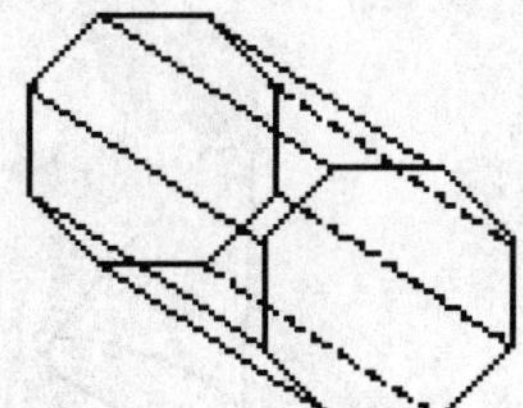

____

9. 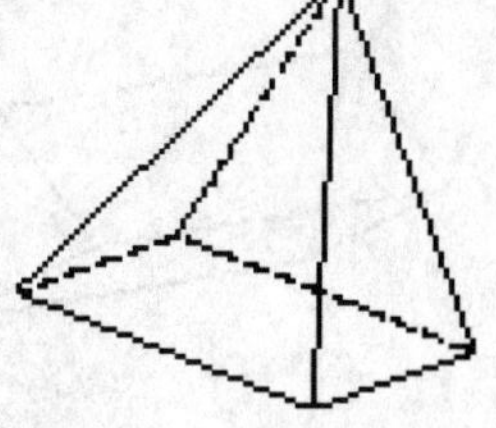

____

Holt Mathematics

LESSON 10-1
Practice B
Introduction to Three-Dimensional Figures

Identify the base or bases of each figure. Then name the figure.

1. 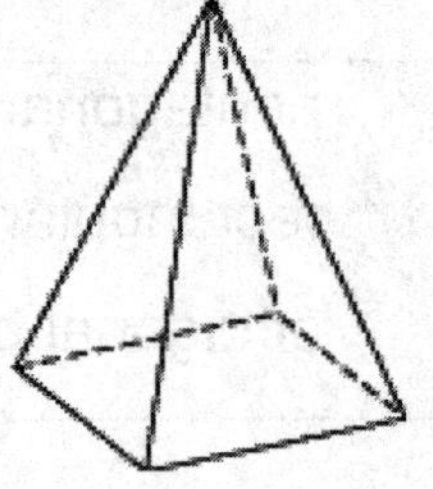

____

____

2. 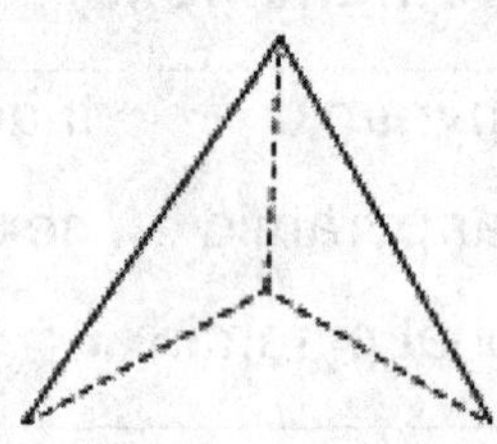

____

____

3. 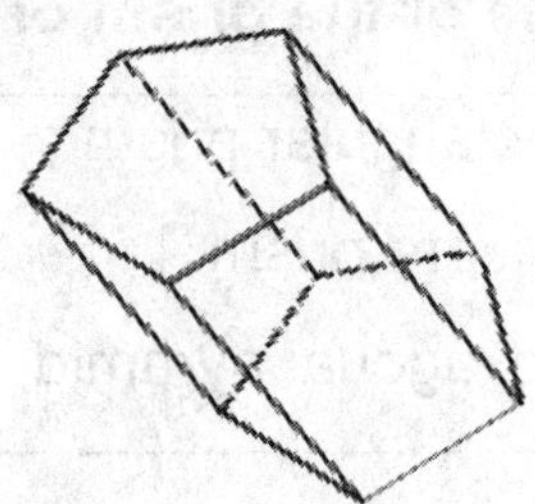

____

____

4. 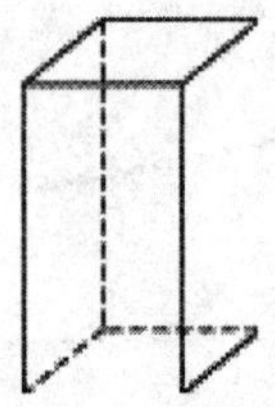

____

____

5. 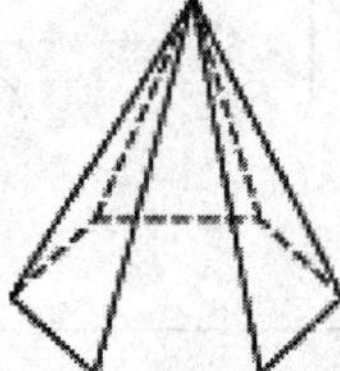

____

____

6. 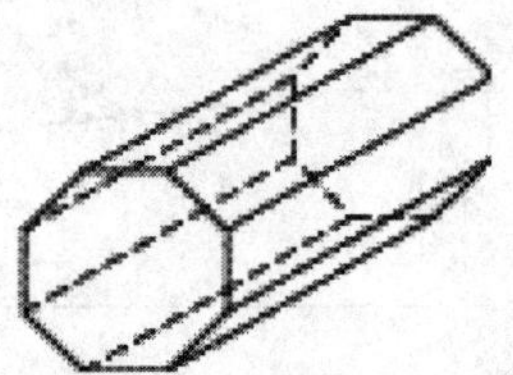

____

____

Classify each figure as a polyhedron or not a polyhedron. Then name the figure.

7. 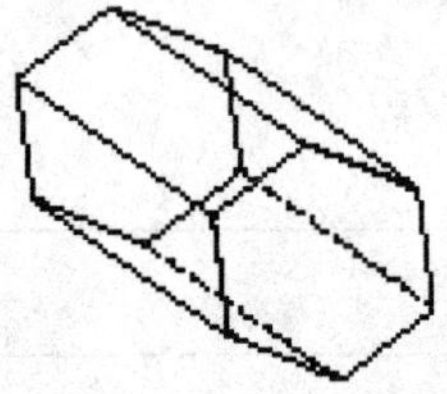

____

____

8. 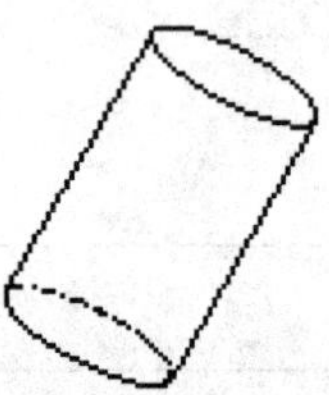

____

____

9. 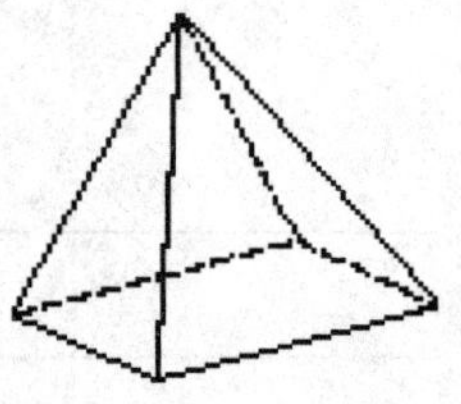

____

____

10. 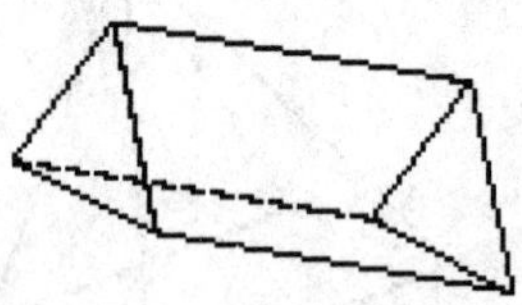

____

____

11. 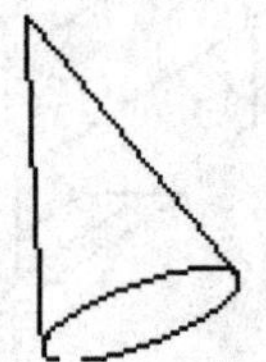

____

____

12. 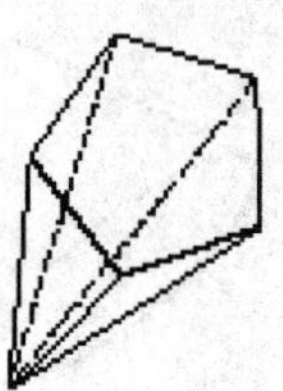

____

____

Holt Mathematics

Practice C
Introduction to Three-Dimensional Figures

Identify the base or bases of each figure. Then name the figure.

1. 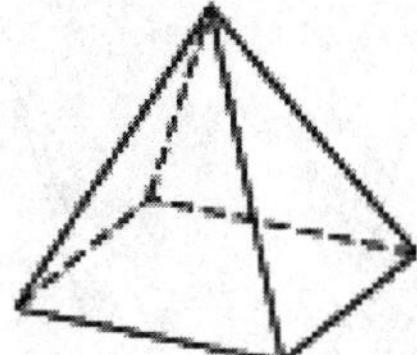

2. 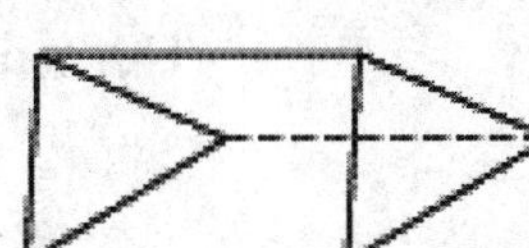

3. 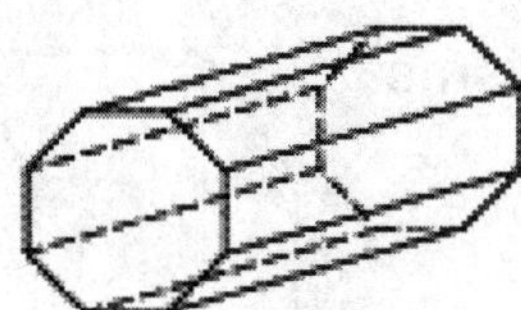

____

____

Classify each figure as a polyhedron or not a polyhedron. Then name the figure.

4. 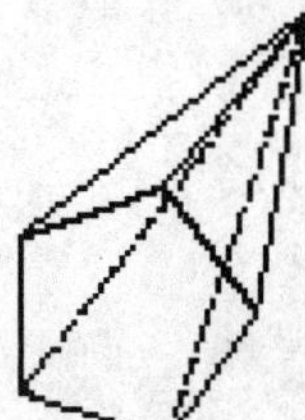

5. 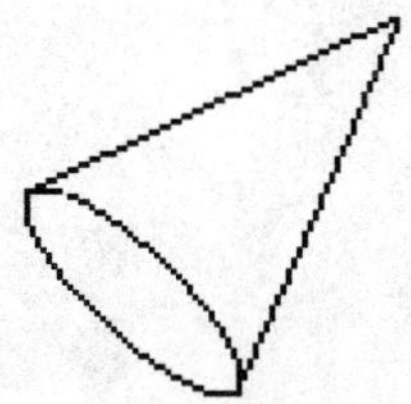

6. 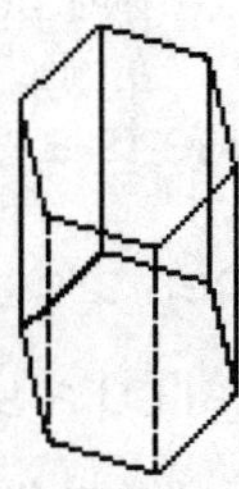

____

____

Identify the three-dimensional figure described.

7. three pairs of congruent rectangular sides _______________________

8. square base, four triangular faces _______________________

9. equilateral triangular base, three congruent equilateral
triangular faces _______________________

10. two pentagonal bases, five parallelogram faces _______________________

11. two congruent circular bases _______________________

12. hexagonal base, six triangular faces _______________________

Holt Mathematics

Reteach
Introduction to Three-Dimensional Figures

A **polyhedron** is a three-dimensional figure whose faces are polygons. There are two types of polyhedrons.

Prisms 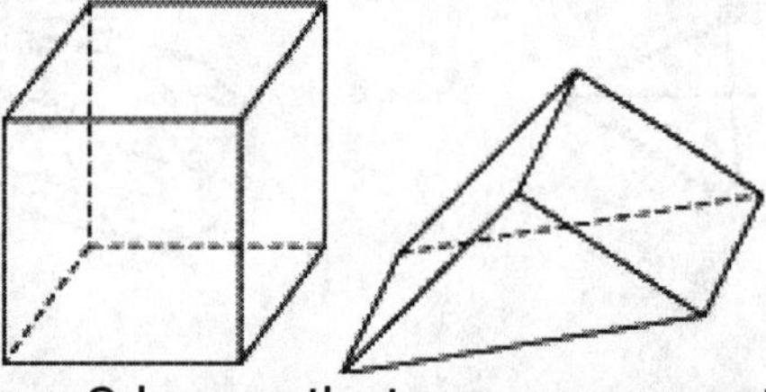

Pyramids 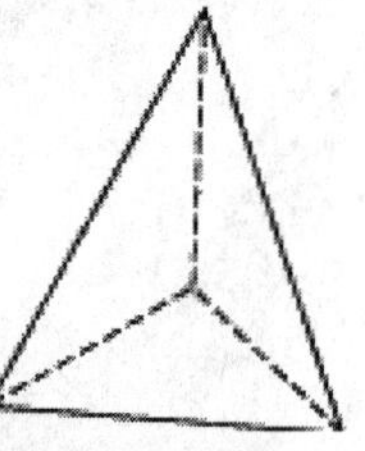

A **prism** has 2 bases that are congruent and parallel polygons. The other faces are parallelograms or rectangles.

A **pyramid** has one base that is a polygon. Its faces are triangles that meet at one vertex.

Prisms and pyramids are named by their bases.

1. Look at the figure at the right.

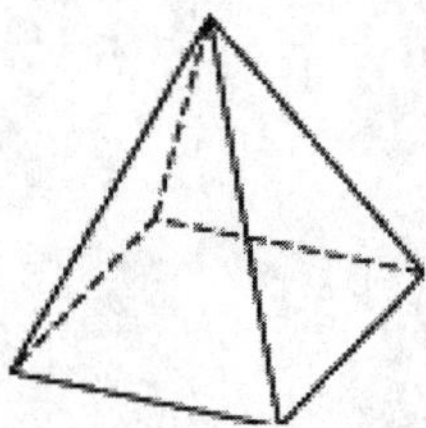

Its base is a _________ _________.

Its faces are _______________.

It is called a square pyramid.

2. Look at the figure at the right.
It has 2 congruent and parallel bases

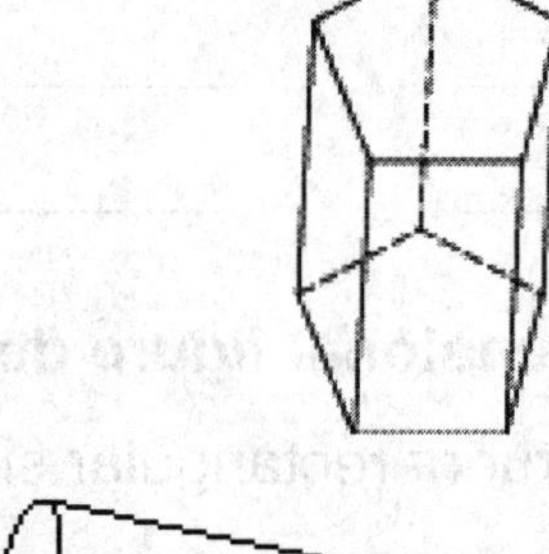

that are _______________.

Its faces are _______________.

It is called a pentagonal prism.

If a three-dimensional figure has a face that is curved, it is not a polyhedron.

not a polyhedron

Is each figure a polyhedron?

3. 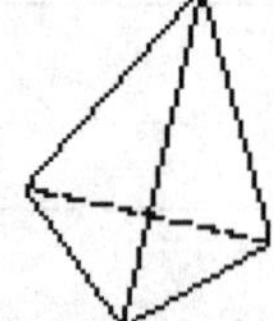

4. 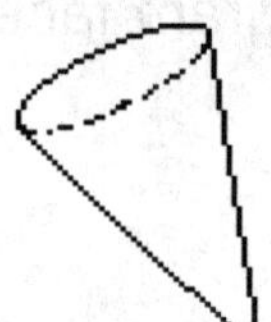

5. 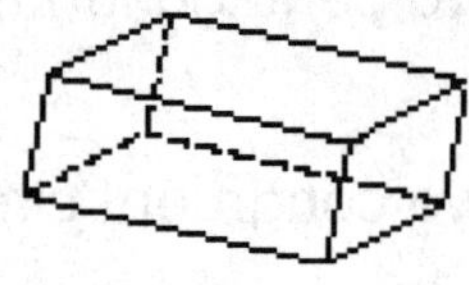

_______________   _______________   _______________

Holt Mathematics

LESSON 10-1 Challenge
Platonic Solids

A three-dimensional figure in which all the faces are polygons is called a *polyhedron*. A polyhedron whose faces are all congruent regular polygons is called a *regular polyhedron*. Regular polyhedrons are called *Platonic solids*.

Below are patterns for four Platonic solids. Copy each pattern. You may enlarge the pattern if you like. Then cut out the pattern and fold it to build a model of the solid. Use your models to answer the questions.

Hexahedron

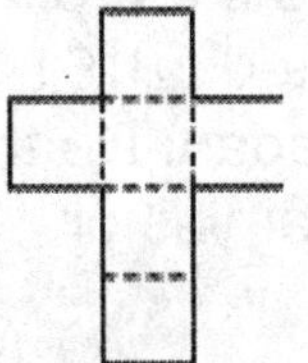

Tetrahedron

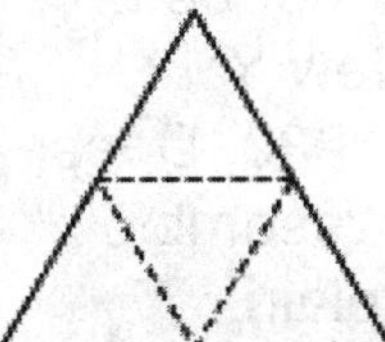

Octahedron

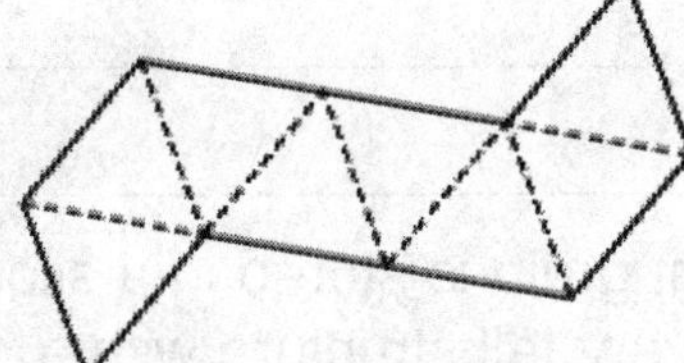

Dodecahedron

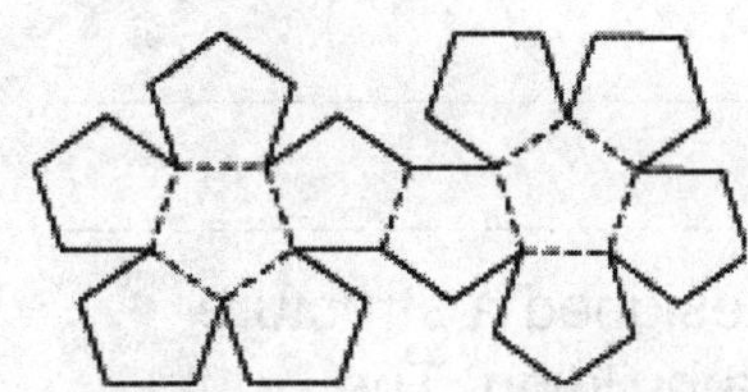

1. How many faces does a hexahedron have? What is another name for a hexahedron?

____

2. How many faces does a tetrahedron have? What is another name for a tetrahedron?

____

3. How many faces does an octahedron have?

____

4. How many faces does a dodecahedron have?

____

Holt Mathematics

Problem Solving
Introduction to Three-Dimensional Figures

Write the correct answer.

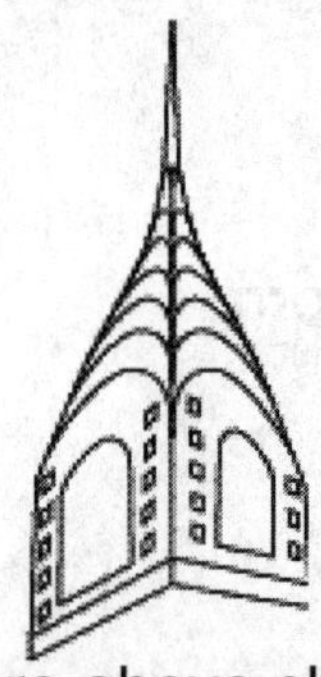

1. The picture above shows the top of the Chrysler Building in New York City. It was completed in 1930. Does the top of the tower most resemble a prism or a pyramid? Explain.

____

____

____

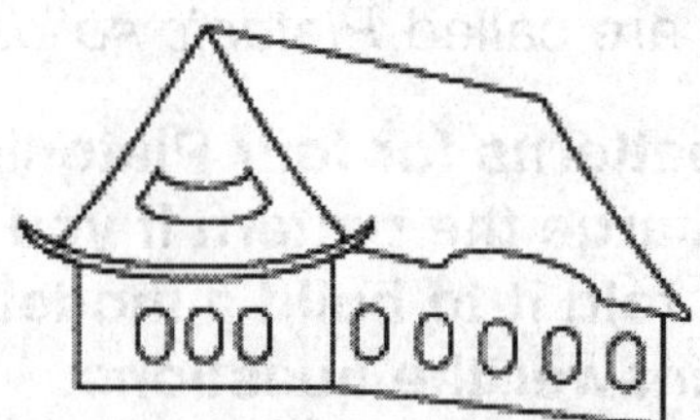

2. The picture above shows the rooftop of Himeji Castle, completed in 1614 in Donjon, Himeji City, Japan. Do the rooftops resemble pyramids or prisms? Explain.

____

____

____

3. An architect designed a structure for the top of a building. The structure has a vertex, one circular base, and a curved surface. What three-dimensional figure is it?

____

4. On a farm, grain is stored in a silo. This is a very tall structure with a circular base and top and a curved surface. What three-dimensional figure does it resemble?

____

Choose the letter of the correct answer.

5. James put two blocks together to build the figure shown. Identify the two figures he used.

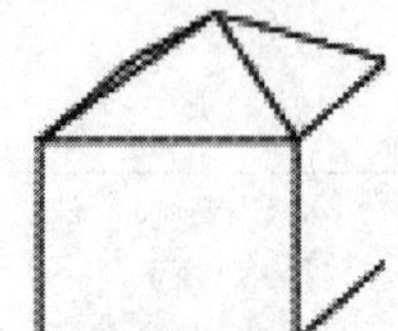

A two pyramids

B a pyramid and a prism

C a pyramid and a cone

D two prisms

6. Jaime constructs a figure that has one rectangular base and three triangular faces. What is the figure?

F a cone

G a triangular pyramid

H a rectangular pyramid

J a rectangular prism

7. The shape of a log is most like which figure?

A cylinder   **C** prism

B pyramid   **D** cone

Holt Mathematics

Reading Strategies
LESSON 10-1 *Focus on Vocabulary*

A **polyhedron** is a three-dimensional figure with flat sides called **faces**.
In a polyhedron, all of the faces are made only of straight sides.

POLYHEDRONS

Prism
- Three-dimensional figure
- Parallel, congruent faces
- Two opposite faces are called "bases."

base
base

Pyramid
- Three-dimensional figure
- Triangular faces meet at a common vertex.
- Face opposite vertex is the base.

vertex
base

1. What name is given to a three-dimensional figure with flat sides? _____________

2. What do you call a flat side of a three-dimensional figure? _____________

3. Which type of polyhedron has two parallel, congruent faces? _____________

4. Which type of polyhedron has a vertex and a face at opposite ends? _____________

5. What do you call the two parallel, congruent faces of a prism? _____________

6. What do you call the face opposite the vertex in a pyramid? _____________

7. Why isn't a cone a polyhedron?

____

8. Name another three-dimensional figure that is not a polyhedron.

____

Holt Mathematics

Puzzles, Twisters & Teasers

LESSON 10-1

You Are Now Entering the X-Word Dimension

Solve the crossword puzzle.

Across

3. A ______ is a three-dimensional figure whose faces are all polygons.

6. A hexagonal ______ has one base that is a hexagon, and six faces that are triangles.

7. a line segment at which two faces of a prism meet

8. a polyhedron with two parallel congruent bases and all other faces being parallelograms

10. The face opposite the ______ is the base of the pyramid.

Down

1. A ______ has one circular face and a curved surface that comes to a point called the vertex.

2. a figure in which two congruent circular faces are connected by a curved surface

4. A ______ prism has six faces that are rectangles.

5. A prism is named for its two congruent parallel ______.

9. Each surface of a polyhedron is called a ______.

Holt Mathematics

LESSON 10-2 Practice A
Volume of Prisms and Cylinders

Find how many cubes each prism holds. Then give the prism's volume.

1. 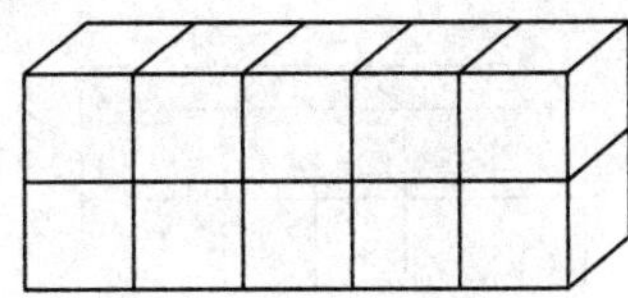

   ____

   ____

2. 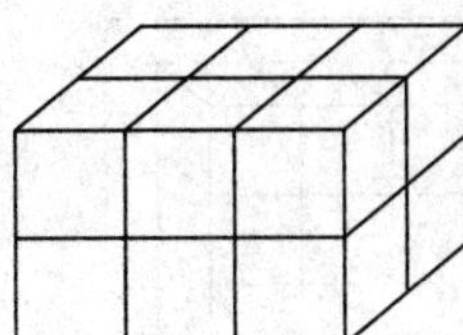

   ____

   ____

3. 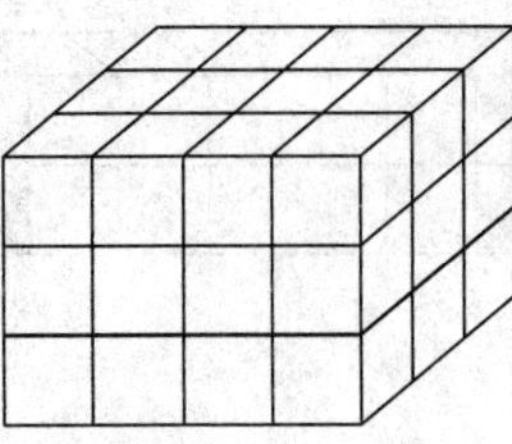

   ____

   ____

Find the volume of each figure. Choose the letter for the best answer.

4. 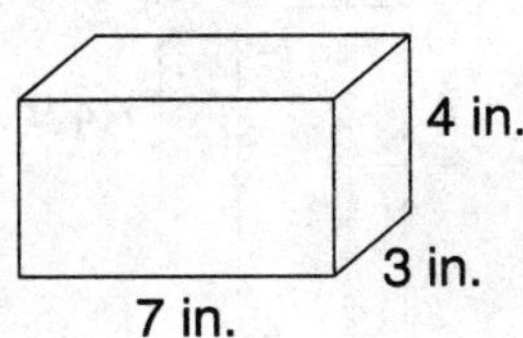

   A 14 in^3 **C** 28 in^3

   B 21 in^3 **D** 84 in^3

5.

   F 40 cm^3 **H** 180 cm^3

   G 72 cm^3 **J** 360 cm^3

6. 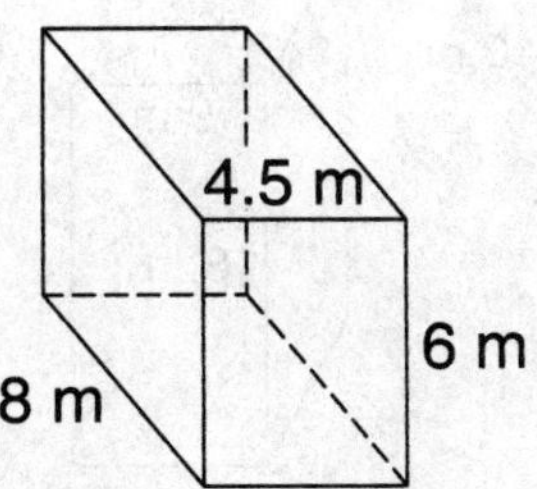

   A 48 m^3 **C** 216 m^3

   B 162 m^3 **D** Not Here

7.

   F 60 cm^3 **H** 720 cm^3

   G 360 cm^3 **J** 810 cm^3

8. A juice can is shaped like a cylinder. It is 12 centimeters wide and 16 centimeters tall. Find its volume to the nearest whole number. Use 3.14 for π.

   ____

Holt Mathematics

LESSON 10-2

Practice B
Volume of Prisms and Cylinders

Find how many cubes each prism holds. Then give the prism's volume.

1.
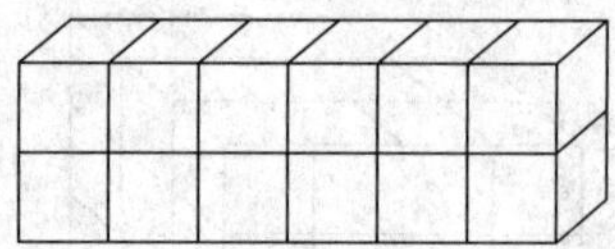
____
____

2.
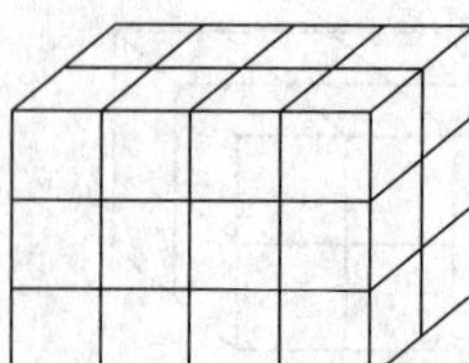
____
____

3.
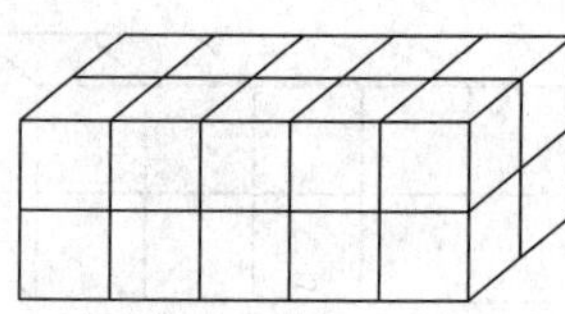
____
____

Find the volume of each figure.

4.
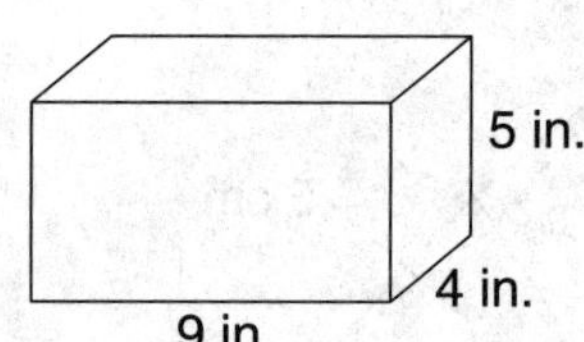

____

5.
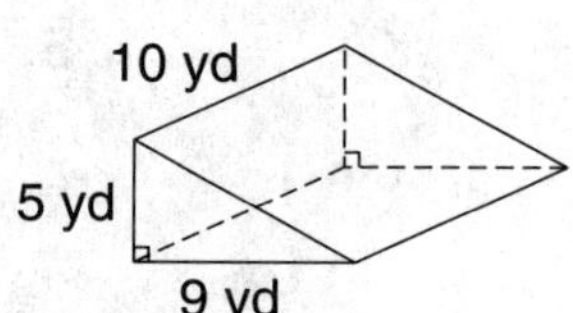

____

6.
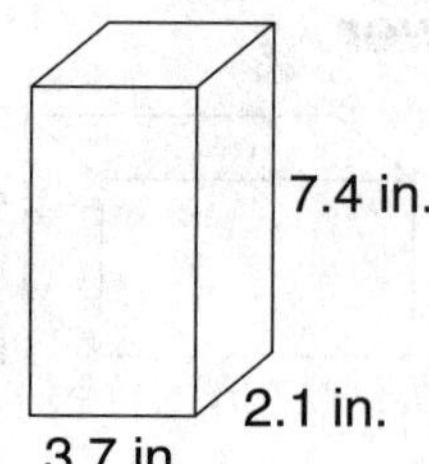

____

7.
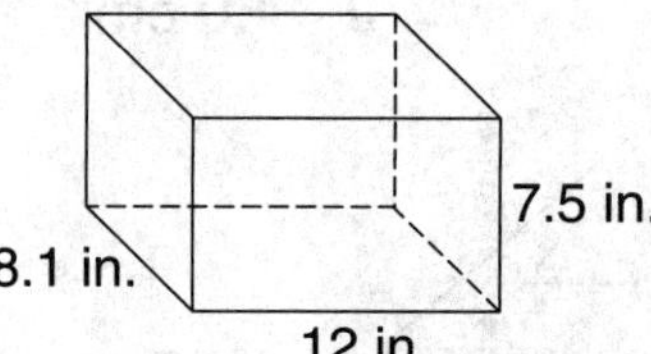

____

8.
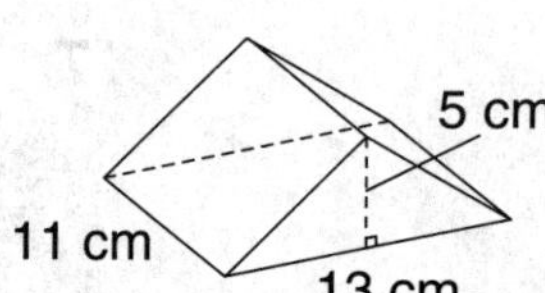

____

9.
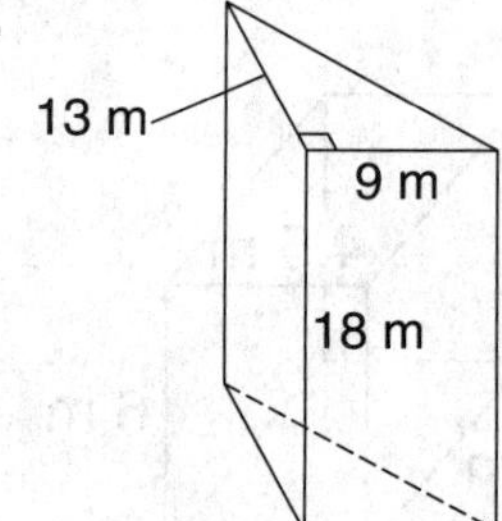

____

10. A travel mug is shaped like a cylinder. It is 9 centimeters wide
and 15 centimeters tall. Find its volume to the nearest tenth.
Use 3.14 for π.

____

Holt Mathematics

Practice C
Volume of Prisms and Cylinders

Find how many cubes each prism holds. Then give the prism's volume.

1. 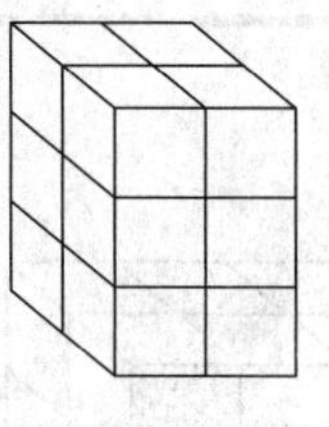

____

____

2. 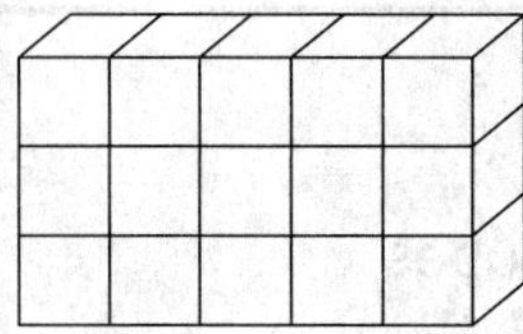

____

____

3. 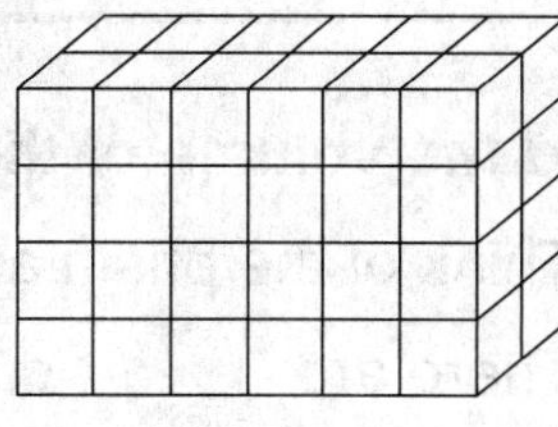

____

____

Find the volume of each figure.

4. 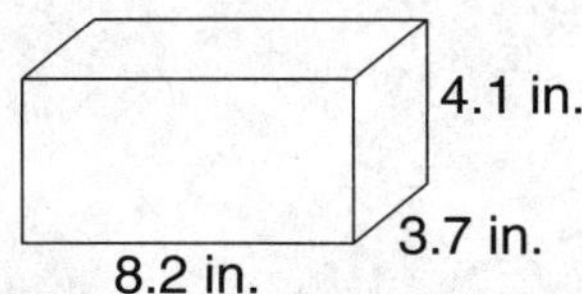

____

5. 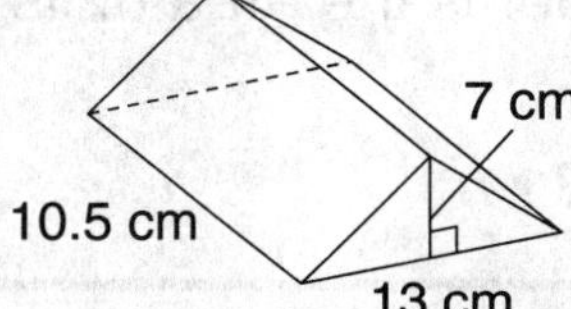

____

6. 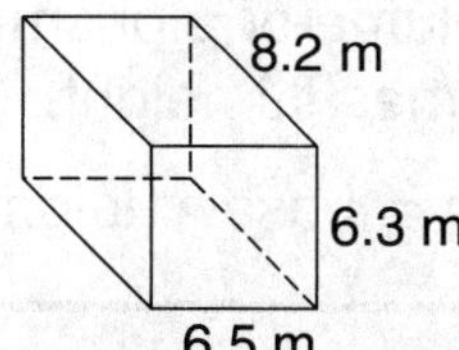

____

7. 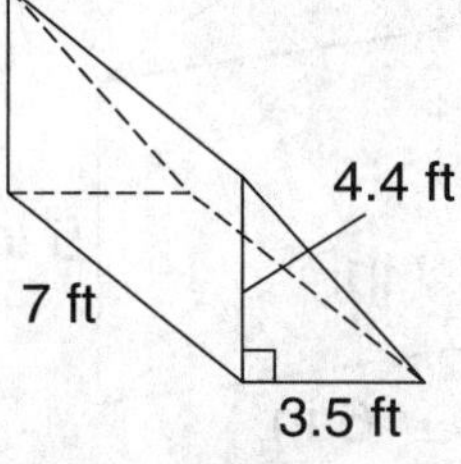

____

8. 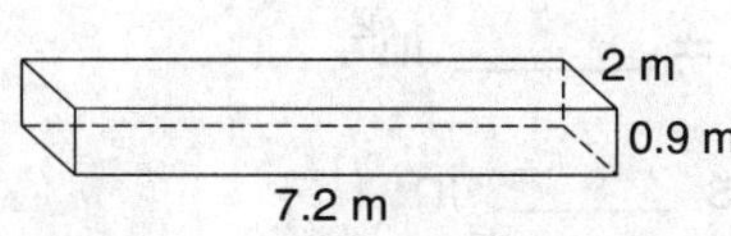

____

9. 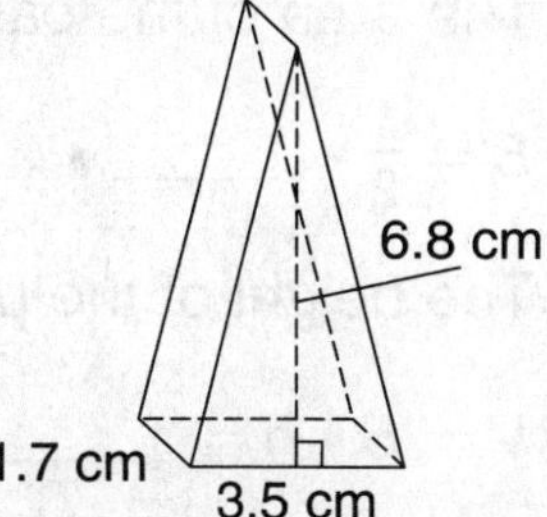

____

10. A vase is in the shape of a cylinder with a diameter of 5 inches and a height of 9 inches. What is the volume of the vase? _______________

11. A glass is shaped like a cylinder. The glass has a volume of 602.88 centimeters and a radius of 4 centimeters. What is the height? _______________

12. The base of a triangular prism is a right triangle with hypotenuse 13 inches long and one leg 12 inches long. The height of the prism is 8 inches. What is the volume of the prism? _______________

Holt Mathematics

LESSON
Reteach

10-2   *Volume of Prisms and Cylinders*

The **volume** of a three-dimensional figure is the amount of space it takes up. Volume is measured in cubic units.

Find the volume of the prism.

1. Think of the prism as layers of cubes.

   There are _______ cubes in the bottom layer.

2. There are _______ layers of cubes.

3. Multiply the number of cubes in the bottom layer by the number of layers.

   The volume of the prism is _______ cubic units.

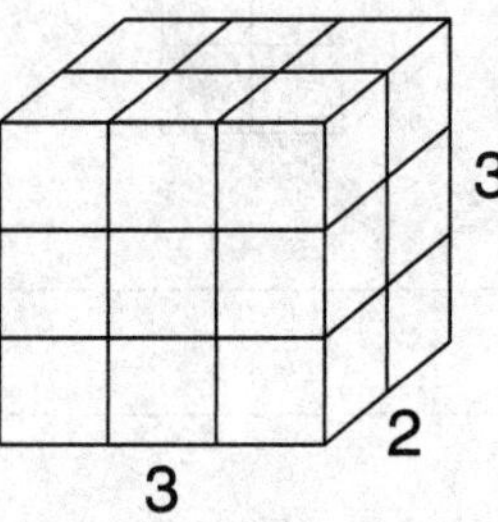

The volume of a prism or a cylinder is the area of its base times its height.

volume = base • height, or $V = B \cdot h$

Find the volume of the prism.

4. What is the shape of the base? _______________

5. The area of the base is $B = \frac{1}{2}bh$

   $B = \frac{1}{2} \cdot$ _______ $\cdot$ _______ = _______ in^2

6. The height of the prism is _______ in.

7. $V = B \cdot h =$ _______ $\cdot$ _______ = _______ in^3

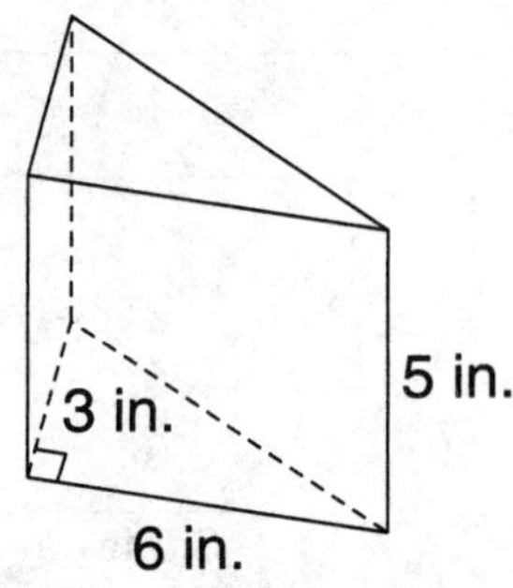

Find the volume of the cylinder to the nearest whole number.

8. What is the shape of the base? _______________

9. The area of the base is $A = \pi r^2$.

   $A = 3.14 \cdot$ _______^{2} = _______ cm^2

10. The height of the cylinder is _______ cm.

11. $V = B \cdot h =$ _______ $\cdot$ _______ = _______ cm^3

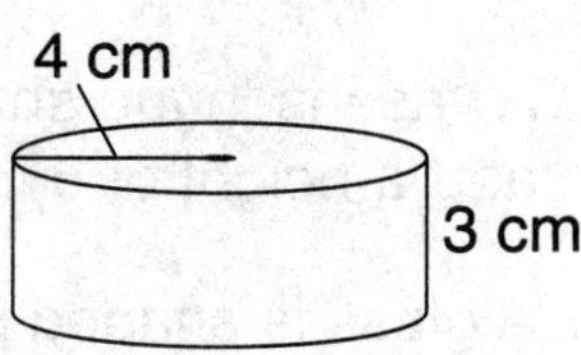

Holt Mathematics

Challenge
LESSON 10-2 *Painted Faces*

A cube has six sides, or faces.
Each of the six faces is a square.

This cube measures 2 units on each edge. The faces
of the cube are painted.

1. How many small cubes
make up the large cube? _______________________

2. How many small cubes are painted on 3 faces?
on 2 faces? on 1 face? not at all?

____

This cube measures 3 units on each edge.

3. How many small cubes
make up the large cube? _______________________

4. If the large cube is painted, how many small
cubes will be painted on 3 faces? on 2 faces?
on 1 face? not at all?

____

This cube measures 4 units on each edge.

5. How many small cubes
make up the large cube? _______________________

6. If the large cube is painted, how many small
cubes will be painted on 3 faces? on 2 faces?
on 1 face? not at all?

____

7. Complete the table below by entering the number
of small cubes that will be painted on the given
number of faces.

Size of Large Cube	Painted on 3 Faces	Painted on 2 Faces	Painted on 1 Face	Painted on 0 Faces
2 × 2 × 2				
3 × 3 × 3				
4 × 4 × 4				
5 × 5 × 5				

8. Look at your table in Exercise 7. What is the rule
for the number of cubes painted on 3 faces? _______________

Holt Mathematics

Problem Solving
Volume of Prisms and Cylinders

Write the correct answer.

1. The eight Corinthian columns at the National Building Museum in Washington, DC, are each 75 feet high and 8 feet in diameter. What is the volume of each column?

2. A cubic centimeter holds 1 milliliter of liquid. How many liters of water to the nearest tenth are required to fill a fish tank that is 24 centimeters high, 28 centimeters long, and 36 centimeters wide?

3. There are 231 cubic inches in a gallon. A large juice can has a diameter of 6 inches and a height of 10 inches. How many gallons of juice does the can hold? Round your answer to the nearest tenth.

4. A small gift box that holds a ring is shaped like a cube. The box measures 1.4 inches on each side. What is the volume of the gift box? Round your answer to the nearest tenth.

Choose the letter of the correct answer.

5. The Leaning Tower of Pisa in Italy appears to be cylindrical in shape. It's height is about 56 meters. If the volume of the tower is about 9,891 cubic meters, what is the diameter of the base?

A about 3.5 m   **C** about 15 m

B about 7 m   **D** about 20 m

6. A bricklayer is building a brick rectangular post to anchor a mailbox. The post is 3 feet tall, 2 feet deep, and 2 feet wide. Each brick is 3 inches by 6 inches by 3 inches. How many bricks does he need?

F 12 bricks   **H** 197 bricks

G 54 bricks   **J** 384 bricks

7. The average stone on the lowest level of the Great Pyramid in Egypt was a rectangular prism 5 feet long by 5 feet high by 6 feet deep and weighed 15 tons. What was the volume of the average casing stone?

A 1,800 ft^3 **C** 150 ft^3

B 1,800 ft^2 **D** 150 ft^2

8. A cylindrical barrel is 2.8 feet in diameter and 8 feet high. If a cubic foot holds about 7.5 gallons of liquid, how many gallons of water will this barrel hold?

F about 1,477 gal

G about 369 gal

H about 150 gal

J about 470 gal

Holt Mathematics

<table><tr><td>**LESSON**
10-2</td><td>

Reading Strategies
Use a Visual Aid
</td></tr></table>

Think of the **volume** of a solid figure as the number of cubic units inside the figure.

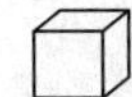

one cubic unit

Each cube represents one cubic unit.

1. What is the height of the prism in the figure? __________________

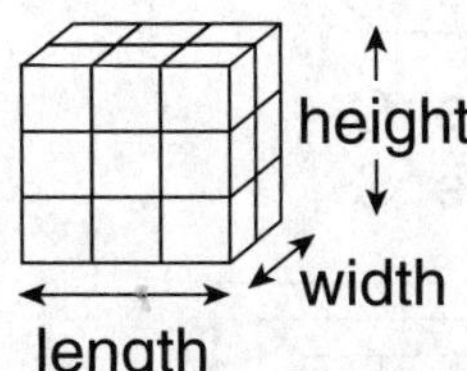

2. What is the width of the prism? __________________

3. What is the length of this prism? __________________

Multiply the length, width, and height of a prism to find its volume in cubic units.

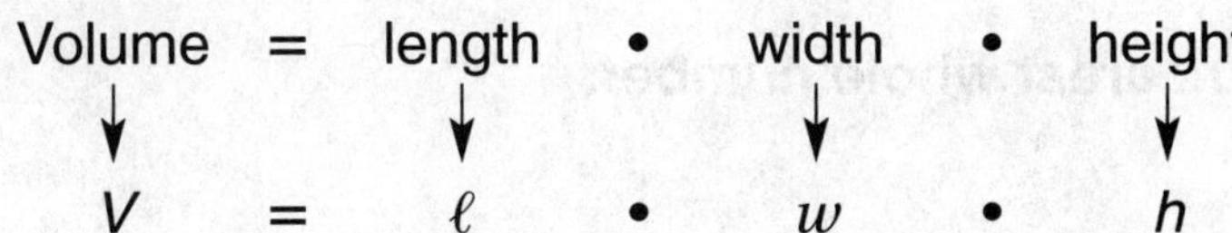

4. Multiply the length, width, and height of the prism above to find the volume. What is the volume? __________________

Use this rectangular prism to complete Exercises 5–7.

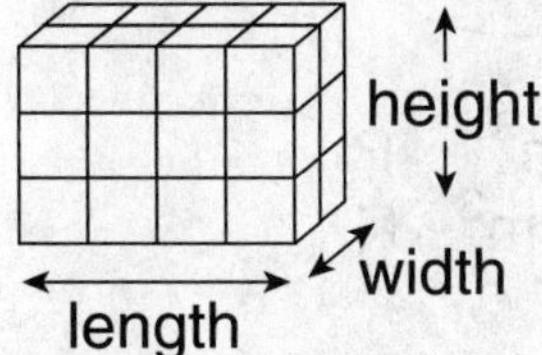

5. How long is the prism? How wide is the prism? __________________

6. What is the height of the prism? __________________

7. Use the formula $V = \ell \cdot w \cdot h$ to find the volume of this prism.

____

Holt Mathematics

Puzzles, Twisters & Teasers

LESSON 10-2

Crack the Code!

Find how many cubes each prism holds. Circle the letter of the correct answer.

1. 16 cubes P
12 cubes L

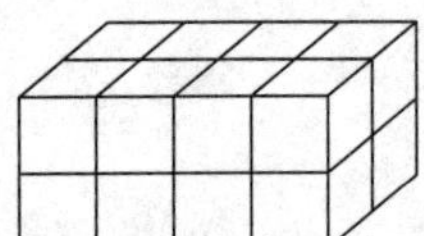

2. 12 cubes Y
16 cubes P

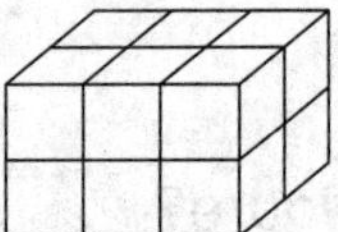

3. 12 cubes Y
24 cubes U

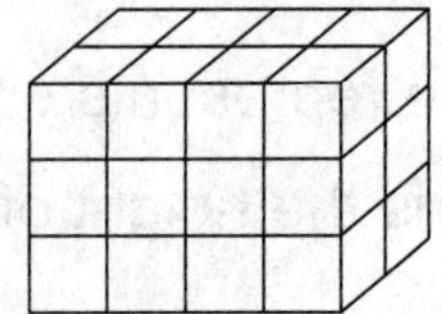

4. 54 cubes A
45 cubes O

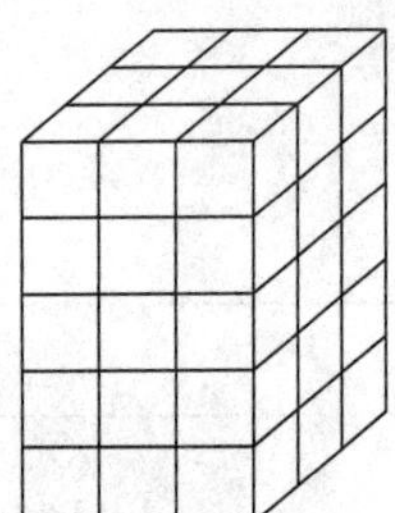

5. 80 cubes E
60 cubes I

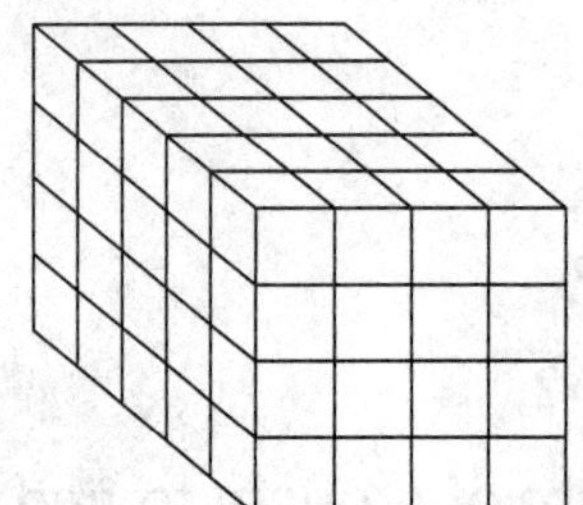

6. 20 cubes C
16 cubes S

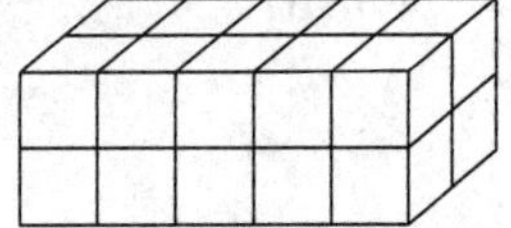

Find the volume of each figure to the nearest whole number. Circle the letter of the correct answer.

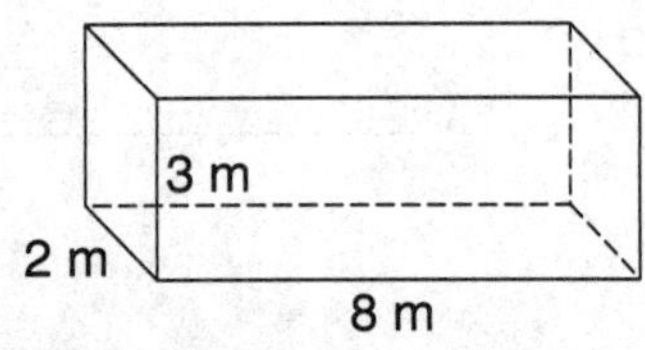

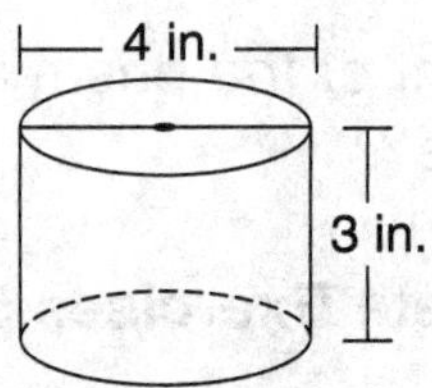

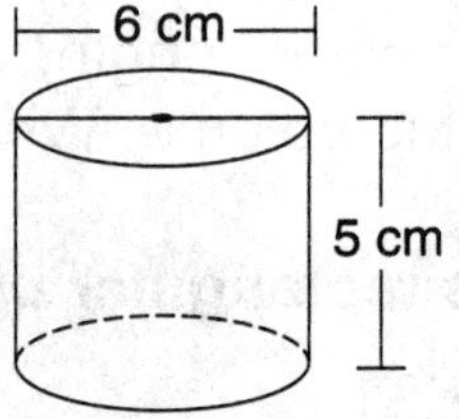

7. 46 m^3 N
48 m^3 M

8. 38 in^3 R
151 in^3 H

9. 47 cm^3 P
141 cm^3 K

Write the letters above the answers to solve the riddle.

What did the egg say to the clown?

___ ___ ___ ___ ___ **A** ___ ___
12 45 24   20 38   20 141

___ ___ ___ ___.
48 80   24 16

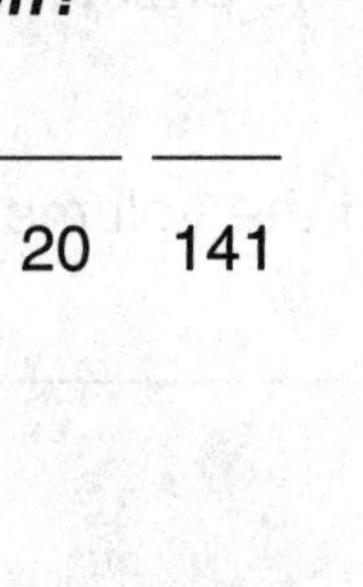

18

Holt Mathematics

LESSON 10-3 — Practice A
Volume of Pyramids and Cones

1. Write the formula for the volume of a pyramid. _______________

2. Write the formula for the volume of a cone. _______________

Find the volume of each pyramid or cone to the nearest whole number. Use 3.14 for π. Cross out each number in the box that matches a volume.

2,826	34	8	500	360	36
471	6	33	7	38	540

3. 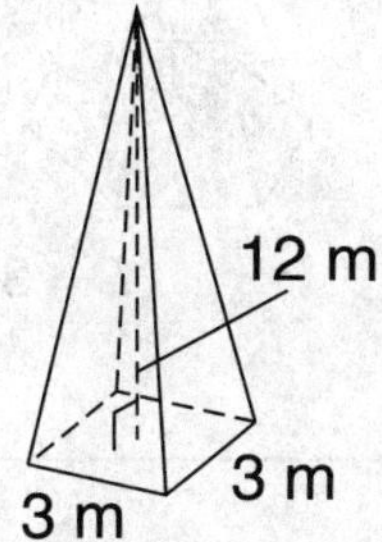

____

4. 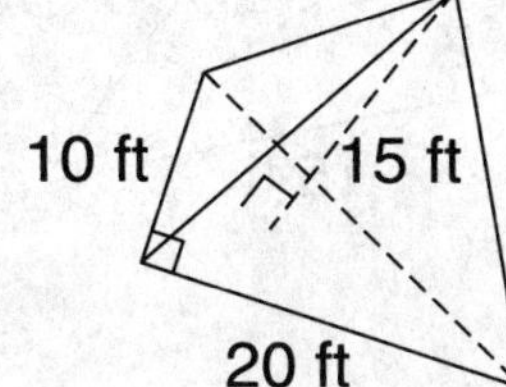

____

5. 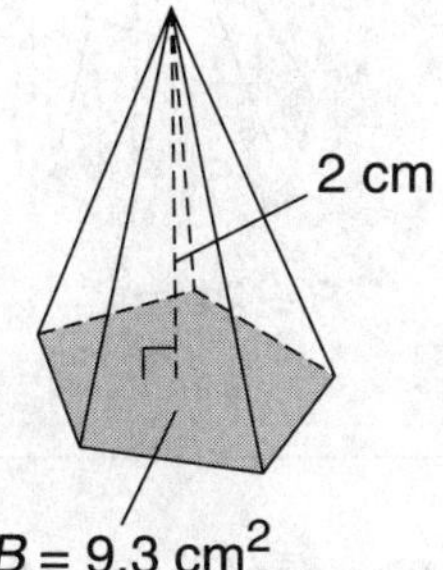

____

6. 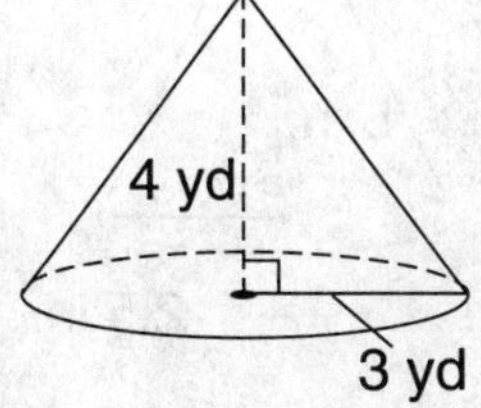

____

7. 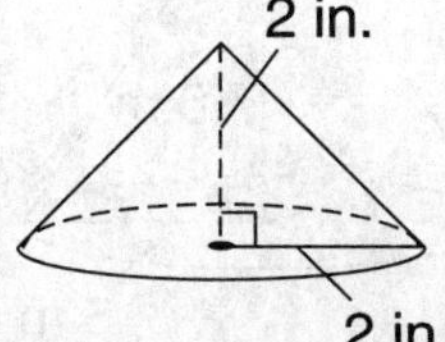

____

8. 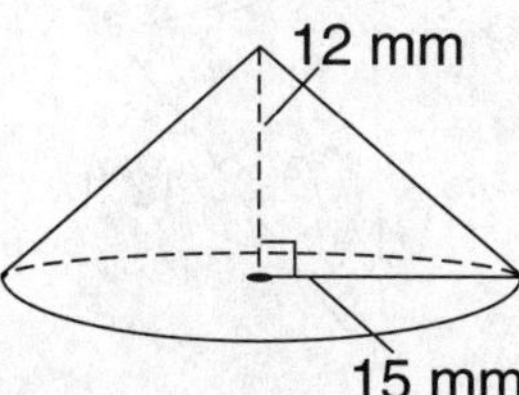

____

9. 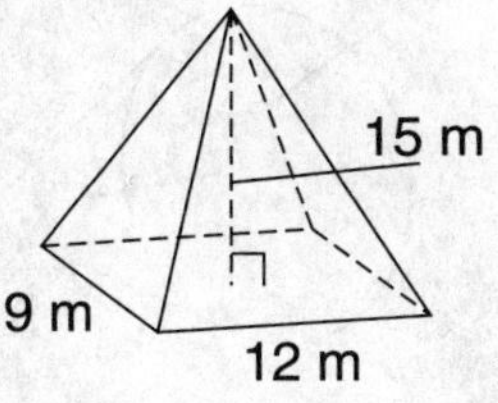

____

10. 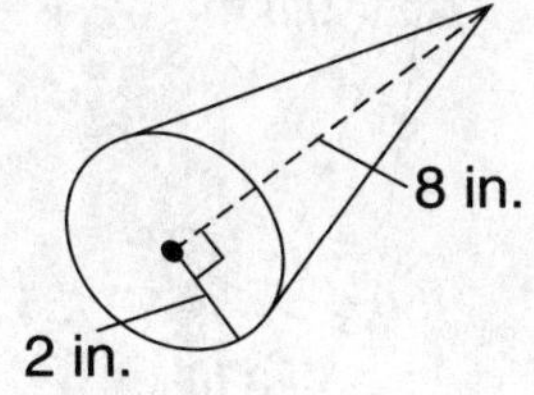

____

11. 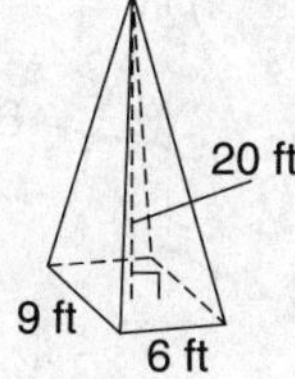

____

Holt Mathematics

Practice B
Volume of Pyramids and Cones

Find the volume of each pyramid to the nearest tenth.

1. 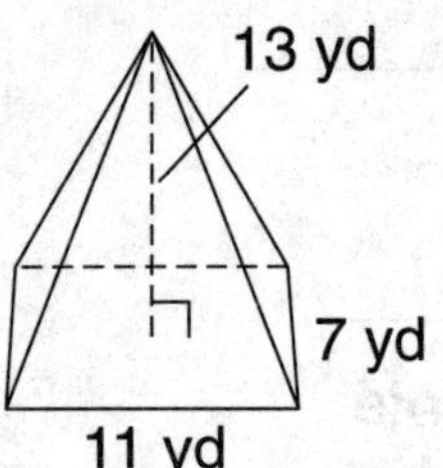

____

2. 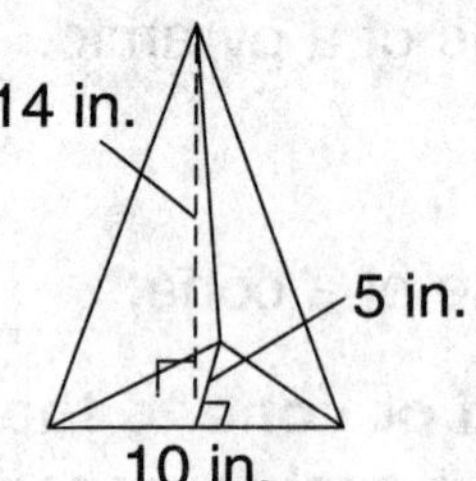

____

3. 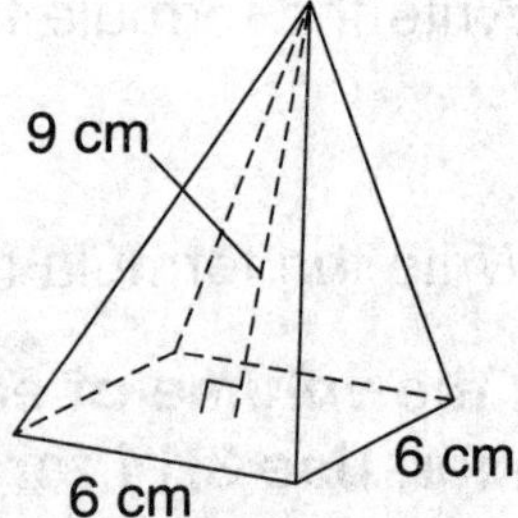

____

4. 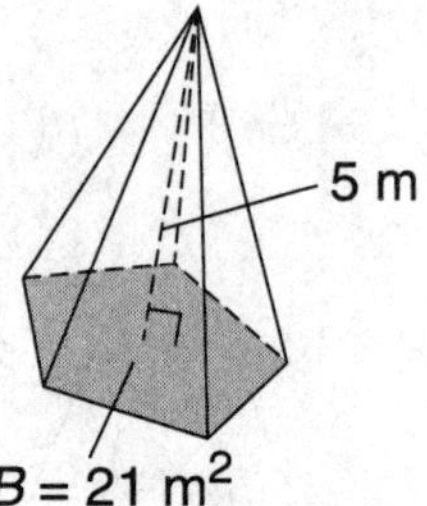

____

5. 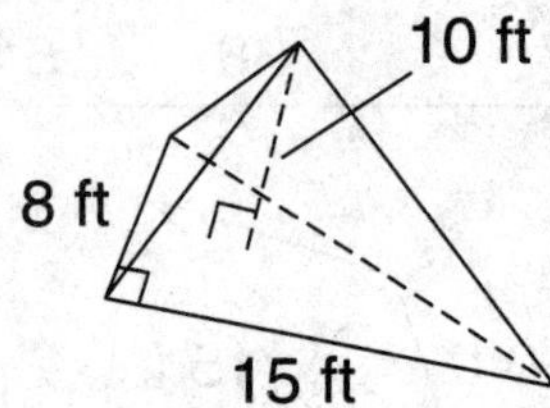

____

6. 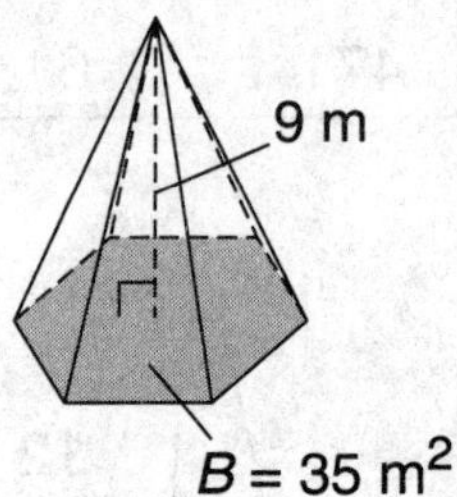

____

Find the volume of each cone to the nearest tenth.

7. 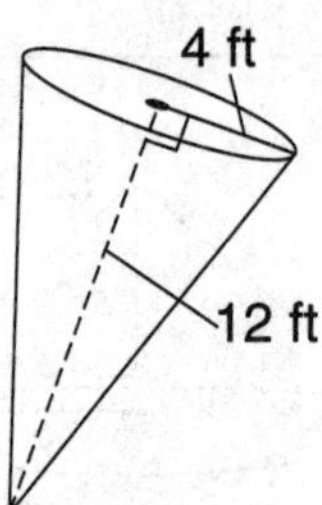

____

8. 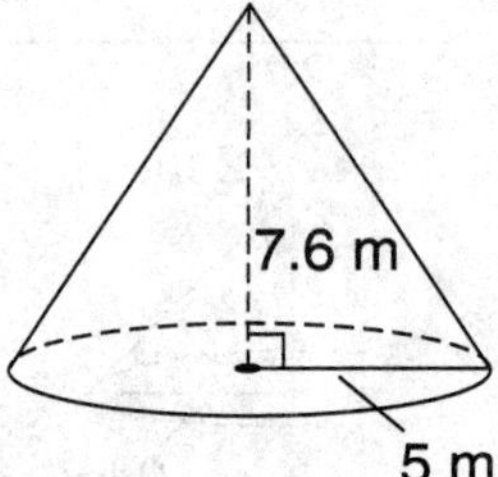

____

9. 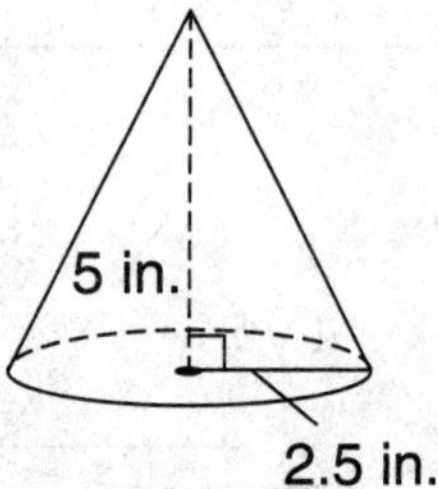

____

10. 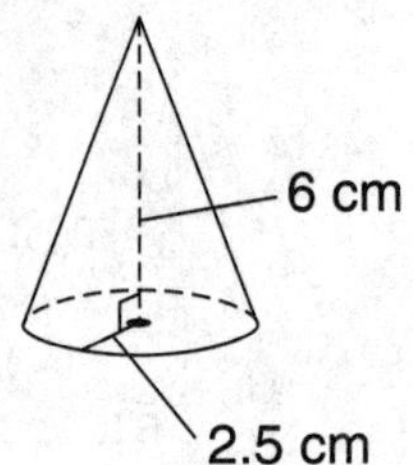

____

11. 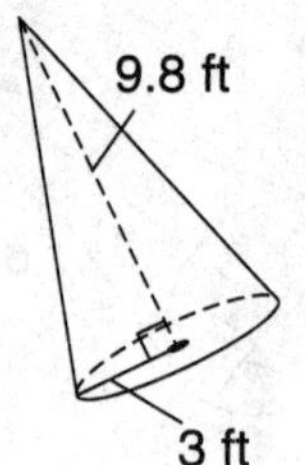

____

12. 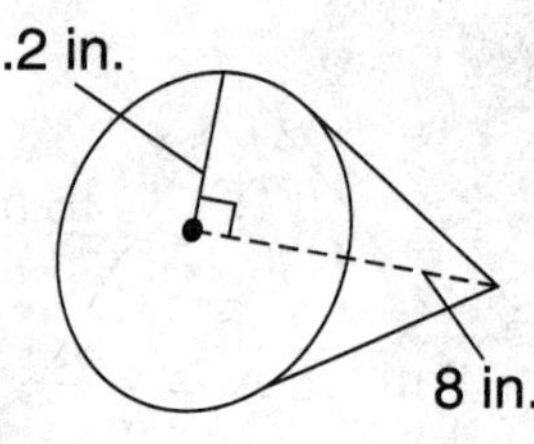

____

Holt Mathematics

Practice C

Volume of Pyramids and Cones

Find the volume of each pyramid to the nearest tenth.

1. 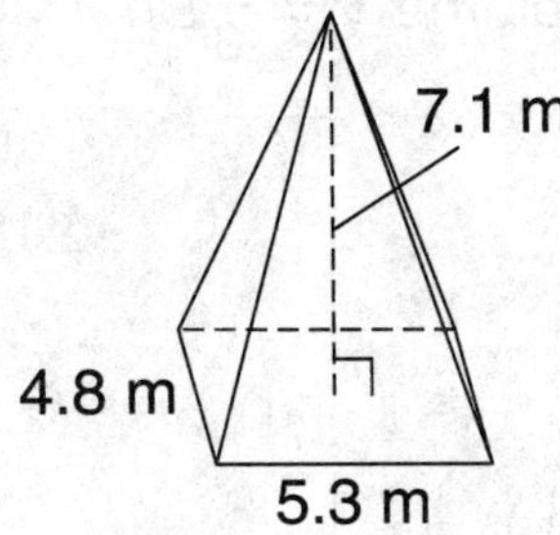

2. 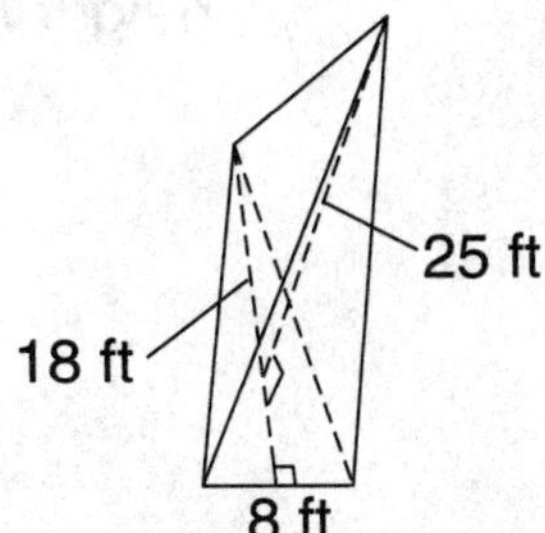

3. 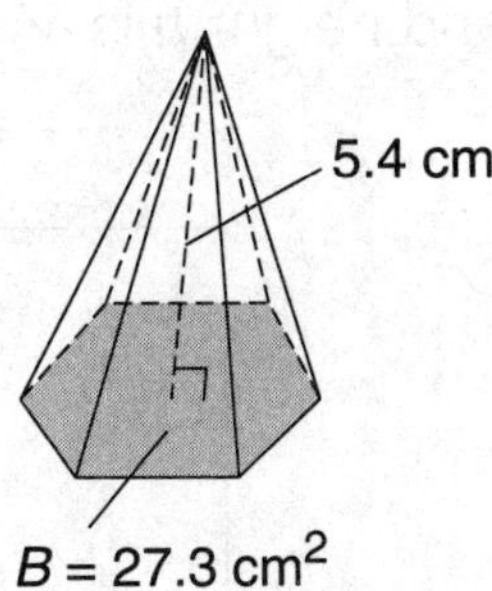

___________   ___________   ___________

Find the volume of each cone to the nearest tenth.

4. 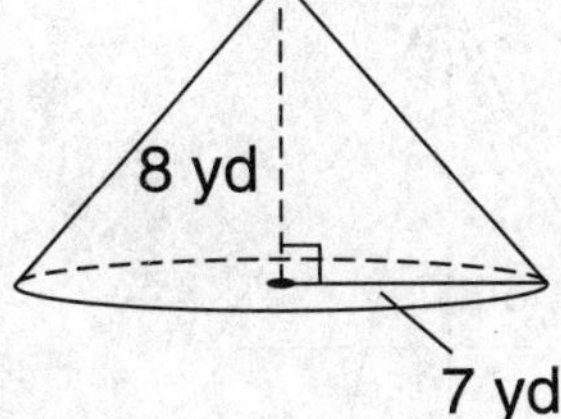

5. 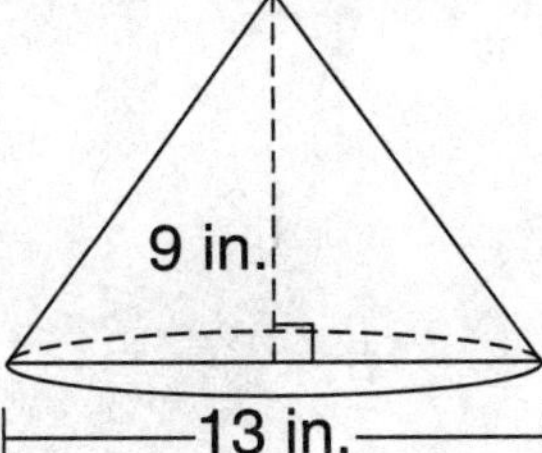

6. 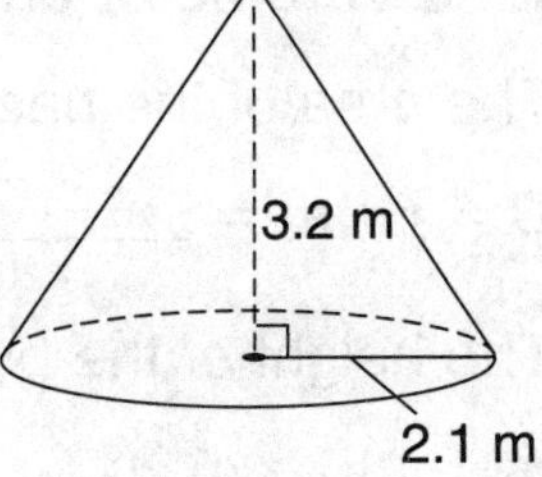

___________   ___________   ___________

Find the volume of each solid to the nearest tenth.

7. rectangular pyramid 7 ft by 3 ft by 9 ft high   ___________

8. cone with radius 8 m and height 11 m   ___________

9. square pyramid with a 4 ft base and height 7 ft   ___________

10. cone with diameter 15 in. and height 20 in.   ___________

11. cone with diameter 11 in. and height of 6 in.   ___________

12. triangular pyramid with height 10 cm and a base
that is a right triangle with legs of 3 cm and 4 cm   ___________

Holt Mathematics

LESSON 10-3 Reteach
Volume of Pyramids and Cones

The volume of a prism with base area B and height h is $V = Bh$.

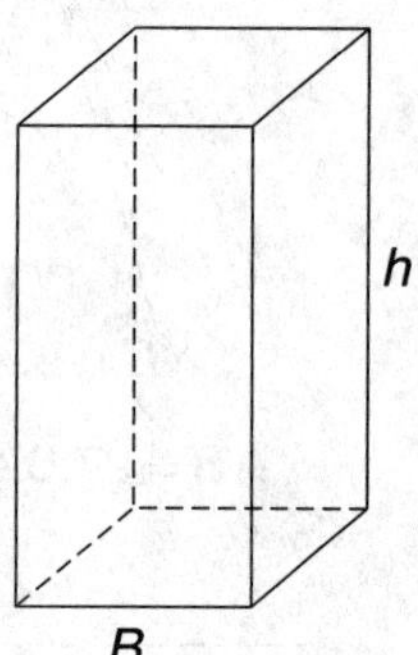

A pyramid with the same base B and height h has the volume $V = \frac{1}{3}Bh$.

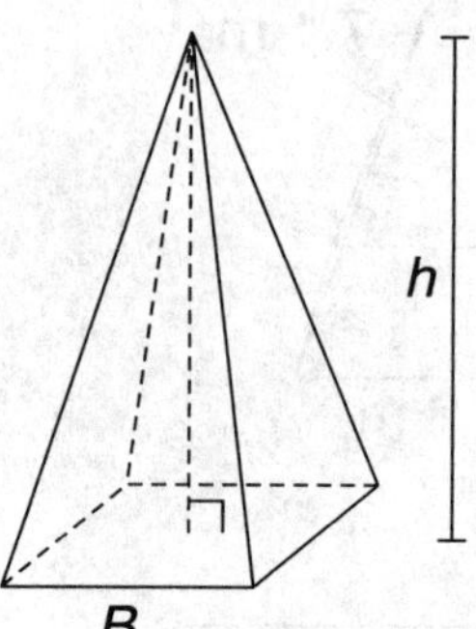

Find the volume of each pyramid.

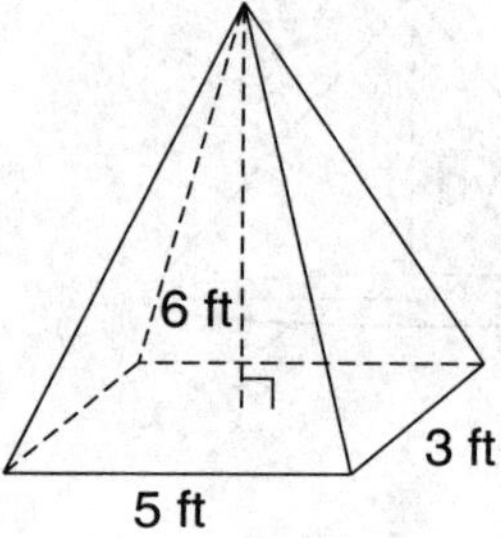

1. The area of the base is

$B = \ell \cdot w =$ _______ $\cdot$ _______ = _______ ft^2

2. The height of the pyramid is _______ ft.

3. $V = \frac{1}{3}Bh = \frac{1}{3} \cdot$ _______ $\cdot$ _______ = _______ ft^3

4. _______________

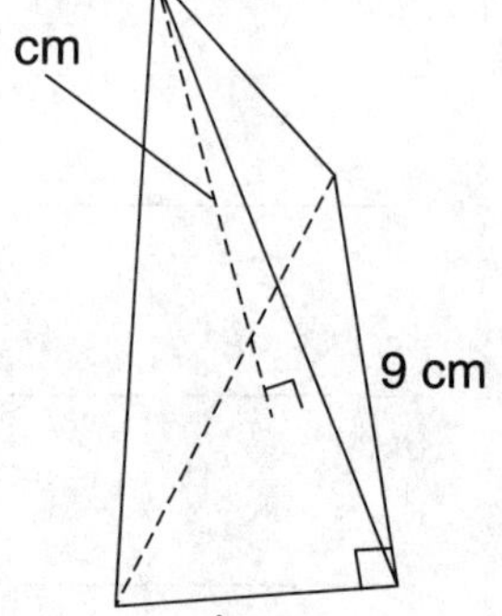

5. _______________

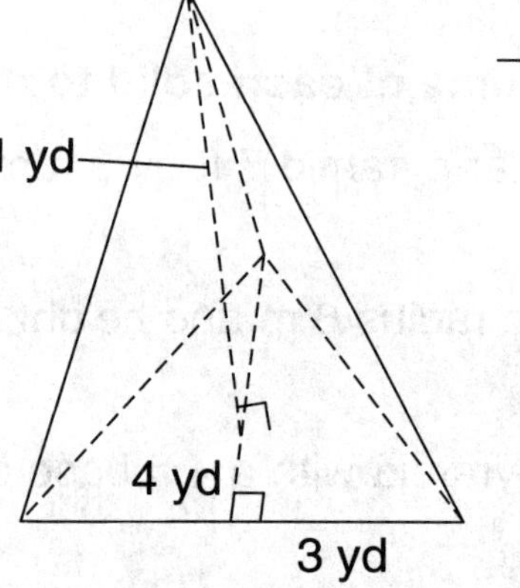

6. _______________

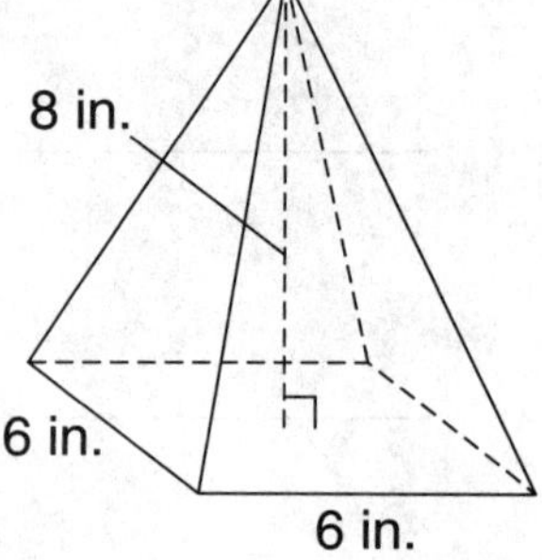

7. _______________

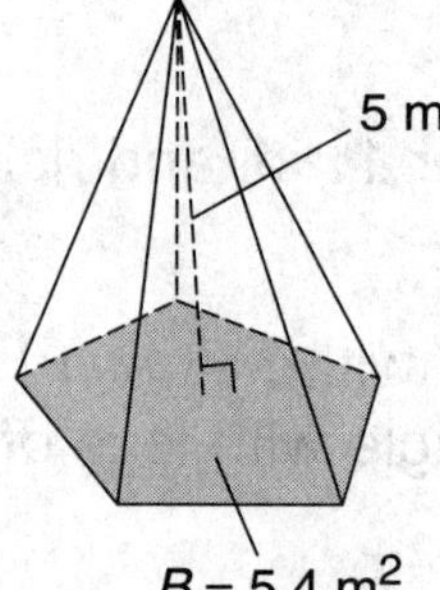

Holt Mathematics

Reteach
Volume of Pyramids and Cones (continued)

You can find the volume of a cone by using the formula $V = \frac{1}{3}Bh$.

Find the volume of each cone to the nearest whole number.

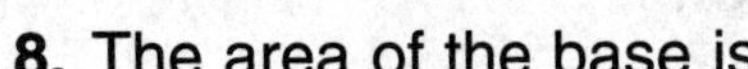

8. The area of the base is

$B = \pi r^2 = 3.14 \cdot \underline{\quad}^2 = \underline{\quad}\ m^2$

9. $V = \frac{1}{3}Bh = \frac{1}{3} \cdot \underline{\quad} \cdot \underline{\quad}$

$ = \frac{1}{3} \cdot \underline{\quad} = \underline{\quad}\ m^3$

10.

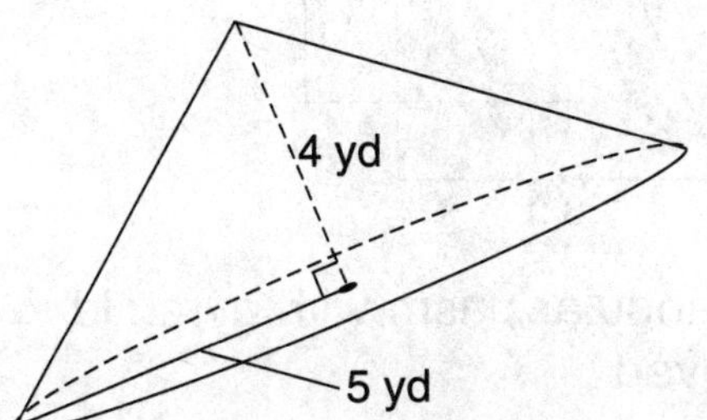

$B = \pi r^2$

$B = 3.14 \cdot \underline{\quad}^2$

$ = 3.14 \cdot \underline{\quad} = \underline{\quad}\ yd^2$

$V = \frac{1}{3}Bh$

$V = \frac{1}{3} \cdot \underline{\quad} \cdot \underline{\quad}$

$ = \frac{1}{3} \cdot \underline{\quad} = \underline{\quad}\ yd^3$

11.

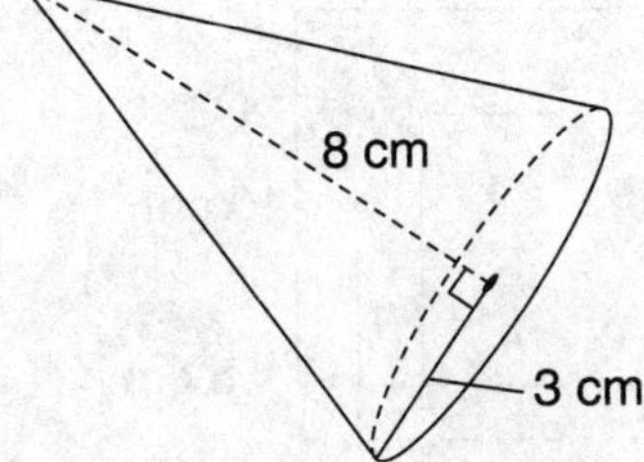

____

12.

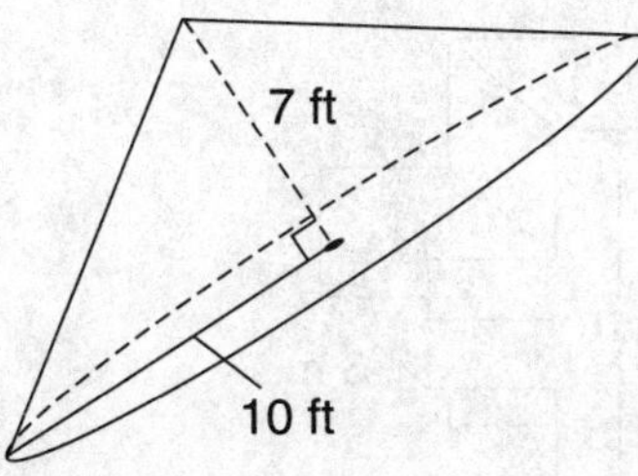

____

13.

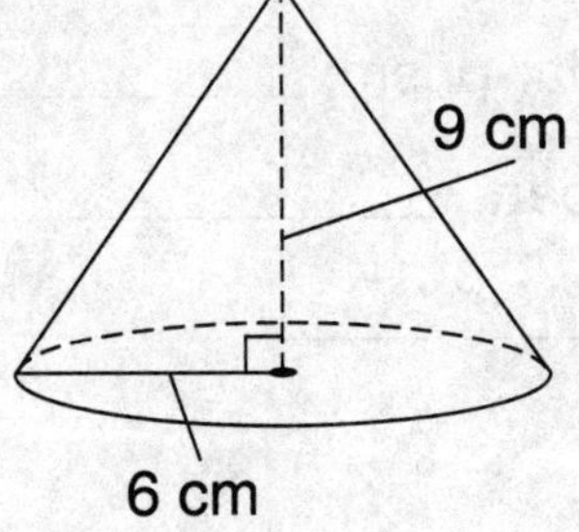

____

14.

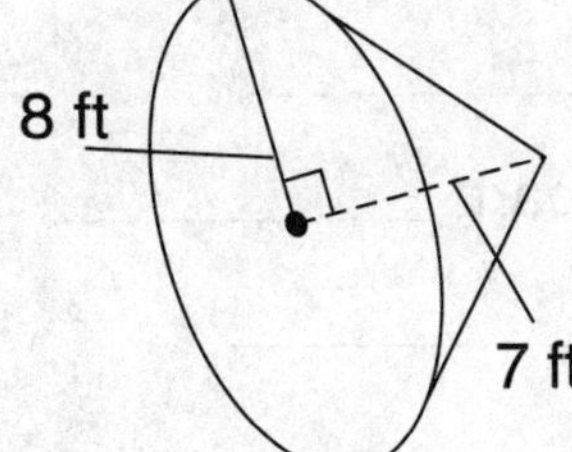

____

Holt Mathematics

LESSON 10-3 Challenge
Take It Out

Sometimes you can find the volume of an oddly shaped figure by first finding the volume of the entire solid and then subtracting a part that is missing.

Find the volume of each figure. Round to the nearest whole number.

1.

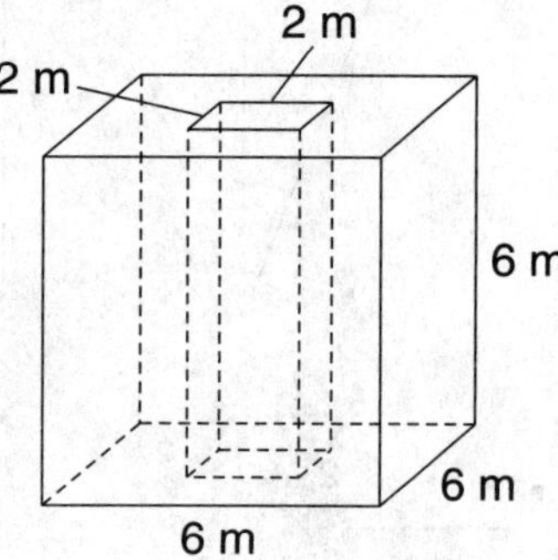

Cube with rectangular prism removed

V of cube: _____________

V of missing part: _____________

V of figure: _____________

2.

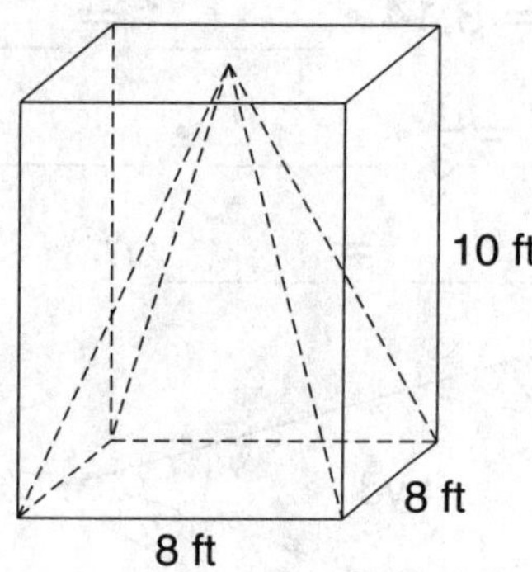

Rectangular prism with pyramid removed

V of prism: _____________

V of missing part: _____________

V of figure: _____________

3.

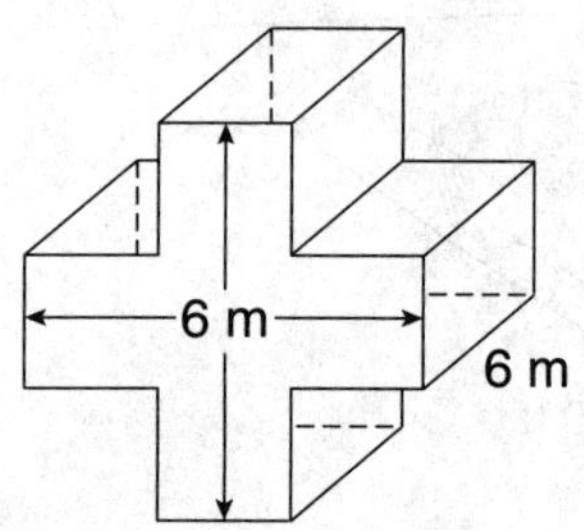

Cube with 2 m × 2 m × 6 m corners removed

V of cube: _____________

V of missing part: _____________

V of figure: _____________

4.

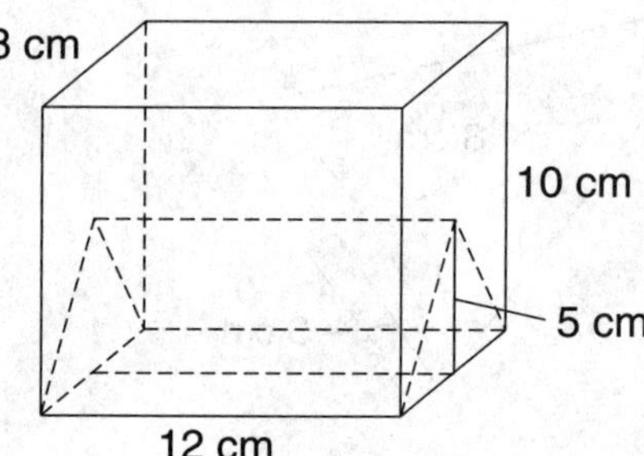

Rectangular prism with triangular prism removed

V of rectangular prism: _____________

V of missing part: _____________

V of figure: _____________

Holt Mathematics

Problem Solving
Volume of Pyramids and Cones

Write the correct answer.

1. Each of the cone-shaped cups near the water cooler has a radius of 3 centimeters and a height of 10 centimeters. If 1 cubic centimeter can hold 1 milliliter of liquid, how much water can each cup hold?

____

2. The Great Pyramid in Egypt has a square base that measures 751 feet on each side. The pyramid is 481 feet high. What is the volume of the Great Pyramid? Round your answer to the nearest cubic foot.

____

3. A waffle cone that holds ice cream is 15 centimeters high and has a diameter of 10 centimeters. What volume of ice cream can it hold if it is filled to the top?

____

4. The base of a rectangular prism is congruent to the base of a pyramid. The height of the pyramid is 3 times the height of the prism. Which figure has a greater volume? Explain.

____

____

Choose the letter of the correct answer.

5. A teepee that is shaped like a cone has a diameter of 12 feet and a height of 15 feet. What is the volume of the teepee?

   A 565.2 ft^3

   B 706.5 ft^3

   C $1,695.6 \text{ ft}^3$

   D $2,119.5 \text{ ft}^3$

6. The top of a 44-story office building is shaped like a pyramid. The base of the pyramid is a right triangle with the two legs measuring 73 feet and 78 feet. The pyramid is 35 feet high. What is the volume of the pyramid?

   F $199,290 \text{ ft}^3$

   G $66,430 \text{ ft}^3$

   H $41,756 \text{ ft}^3$

   J $33,215 \text{ ft}^3$

7. The diameter of a cone-shaped container is 4 inches. Its height is 6 inches. How much greater is the volume of a cylinder-shaped container with the same diameter and height?

   A 50.24 in^3

   B 75.36 in^3

   C 100.98 in^3

   D 200.96 in^3

8. A square pyramid mold for a candle has a base of 64 square centimeters and a height of 12 centimeters. How much greater is the volume of a rectangular prism mold with the same base and height?

   F 64 cm^3

   G 96 cm^3

   H 256 cm^3

   J 512 cm^3

Holt Mathematics

Reading Strategies
LESSON 10-3 *Use a Graphic Organizer*

This chart will help you compare pyramids and cones and the
formulas for finding the volume of these figures.

Pyramid	**Volume of Pyramid**
• Base is a polygon. • Faces are triangles. • Vertex and base are at opposite ends.	Volume = $\frac{1}{3} B \cdot h$ (B = area of base)
Pyramids and Cones	
Cone • Base is a circle. • Has one curved surface. • Vertex is opposite the base.	**Volume of Cone** Volume = $\frac{1}{3}\pi r^2 \cdot h$ (πr^2 = area of base)

Use the graphic organizer to answer each question.

1. Which figure has one curved surface? _________________

2. What forms the base of a pyramid? _________________

3. What forms the base of the cone? _________________

4. What is the formula for the area of a cone's base?

____

5. What shape are the faces of a pyramid?

____

6. Compare the volume formulas. How are they alike?

____

Holt Mathematics

Puzzles, Twisters & Teasers

LESSON 10-3 *Loud and Clear!*

Circle words from the list in the word search. Then find a word that answers the riddle. Circle it and write it on the line.

volume   pyramid   cone   sphere   container

prism   congruent   base   radius   cylinder

```
C O N G R U E N T C V
L U J I A C V B N O W
O O P L D F G H J N C
U D P R I S M E R T Y
D M N B U A S D F A L
P C B A S E T Y U I I
Y W D R T H U I J N N
R W S C D E C O N E D
A T H U Y S P H E R E
M V O L U M E G T R R
I P L M N K O I J B Y
D Q W E R T Y U I I O
```

How does the sky listen to music?

Through a _____________-speaker

Holt Mathematics

 # Practice A
Surface Area of Prisms and Cylinders

Find the surface area of the prism formed by each net. Choose the letter for the best answer.

1.

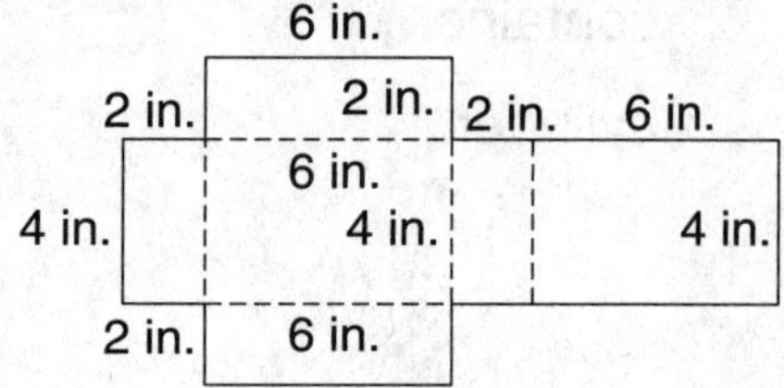

A 36 in^2 **C** 48 in^2

B 44 in^2 **D** 88 in^2

2.

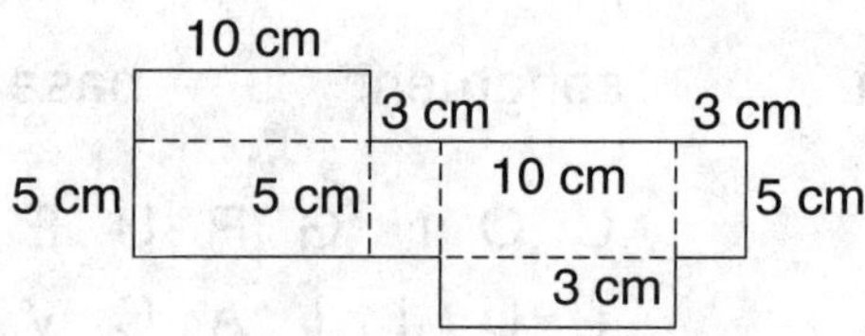

F 95 cm^2 **H** 190 cm^2

G 100 cm^2 **J** 380 cm^2

Find the surface area of the cylinder formed by each net to the nearest whole number. Choose the letter for the best answer.

3.

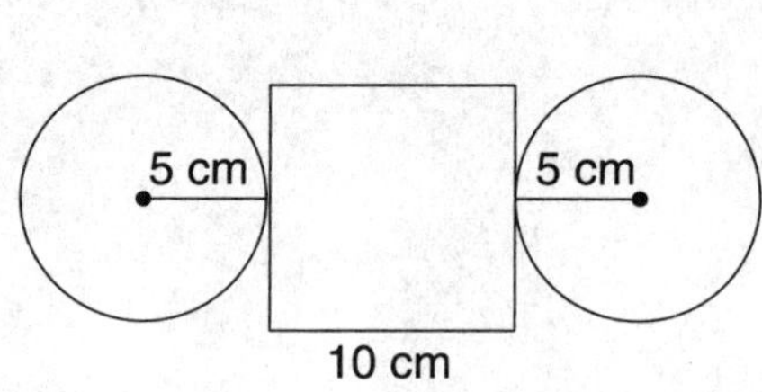

A 157 cm^2 **C** 471 cm^2

B 314 cm^2 **D** 942 cm^2

4.

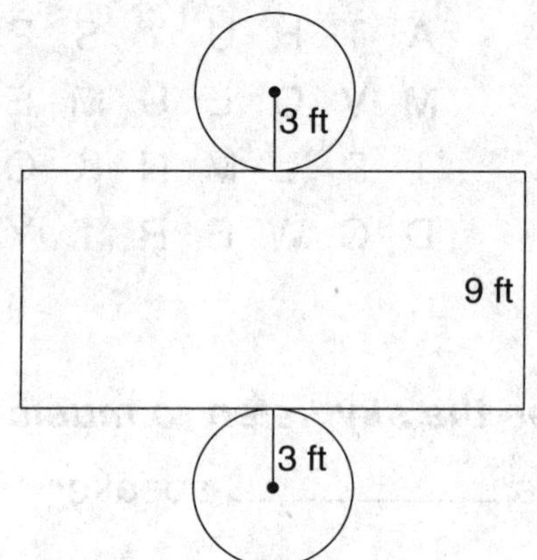

F 226 ft^2 **H** 113 ft^2

G 170 ft^2 **J** 57 ft^2

5. A soup can has a diameter of 8 centimeters and a height of 10 centimeters. The label for the can covers the entire side of the can. About how many square centimeters of paper are needed for the label? Round to the nearest whole number.

____

Holt Mathematics

LESSON 10-4 — Practice B
Surface Area of Prisms and Cylinders

Find the surface area of the prism formed by each net to the nearest tenth.

1.

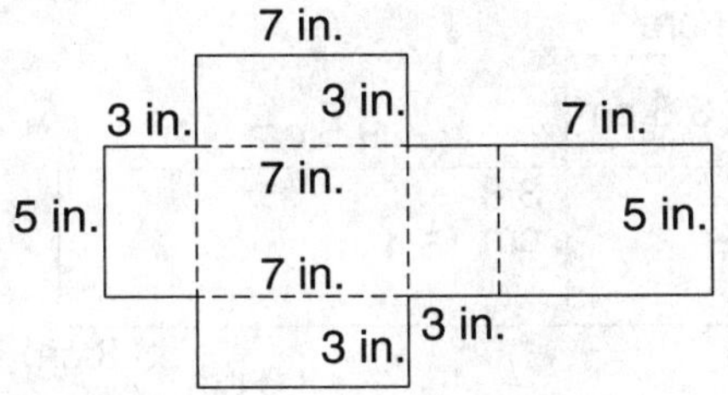

____

2.

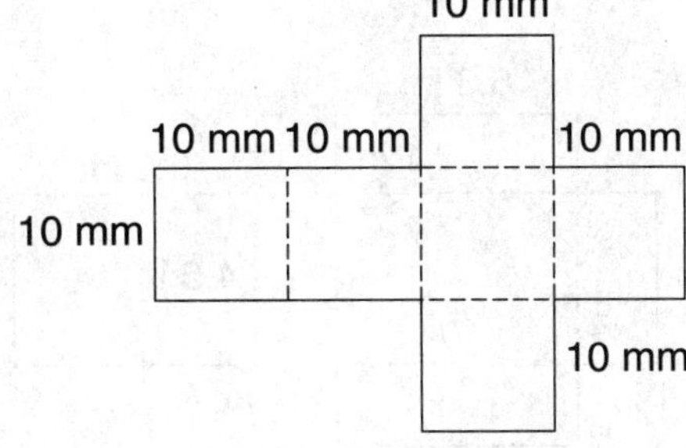

____

Find the surface area of the cylinder formed by each net to the nearest tenth.

3.

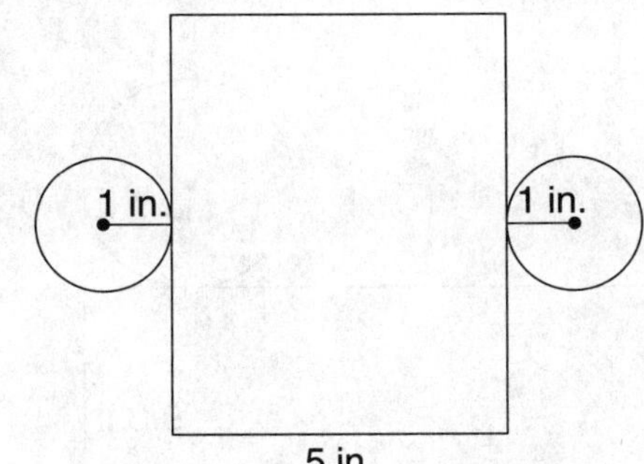

____

4.

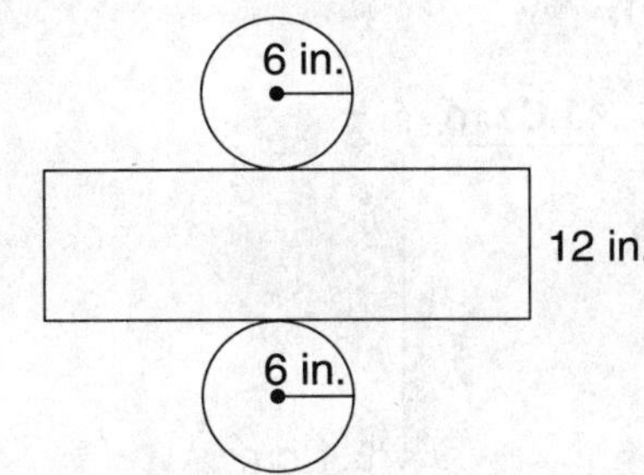

____

5. Mr. Wang has a circular swimming pool with a diameter of 15 feet and a height of 5 feet. Mr. Wang buys a liner to cover the bottom and the sides of the pool. To the nearest square foot, about how many square feet of liner should Mr. Wang buy in order to have enough liner? Explain your answer.

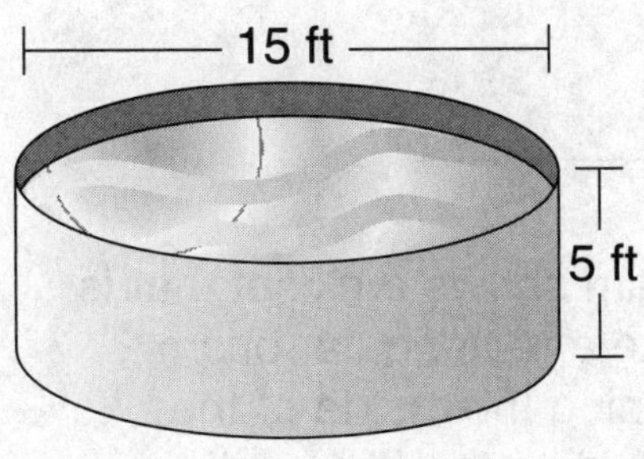

____

____

____

Holt Mathematics

Practice C
Surface Area of Prisms and Cylinders

Find the surface area of the prism formed by each net to the nearest tenth.

1.

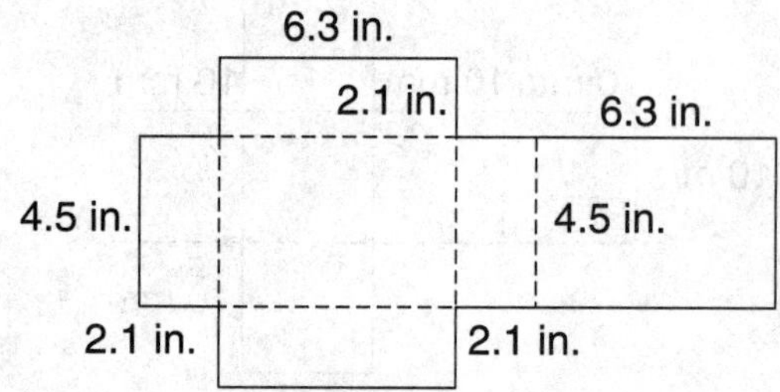

____

2.

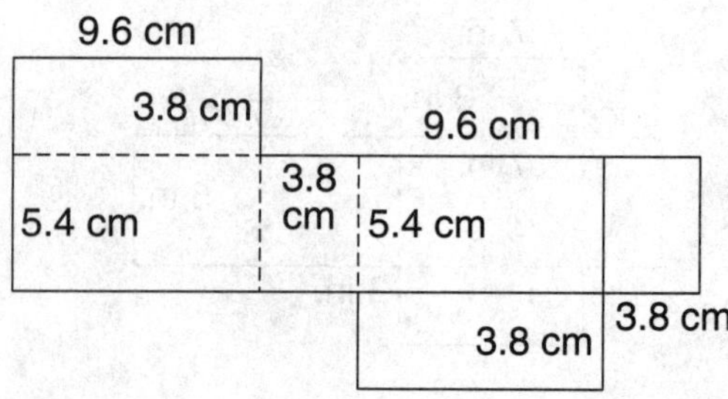

____

Find the surface area of the cylinder formed by each net to the nearest tenth.

3.

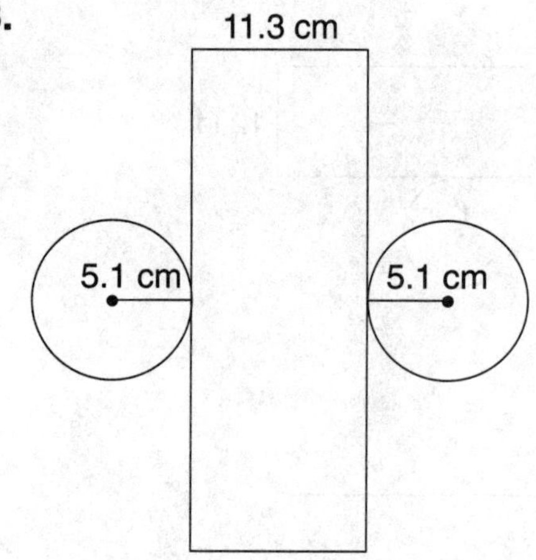

____

4.

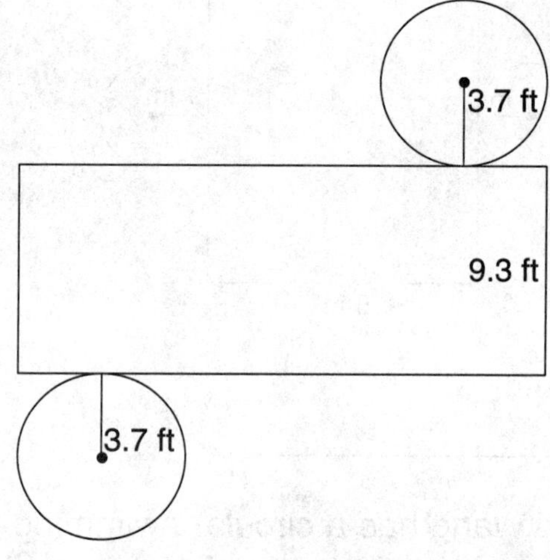

____

5. The diagram shows a closet that is shaped like a rectangular prism. Judith is lining the inside of the closet, including the floor, with cedar. Judith is not lining the doorway. If cedar costs $4.00 per square foot, what will the total cost of the cedar be?

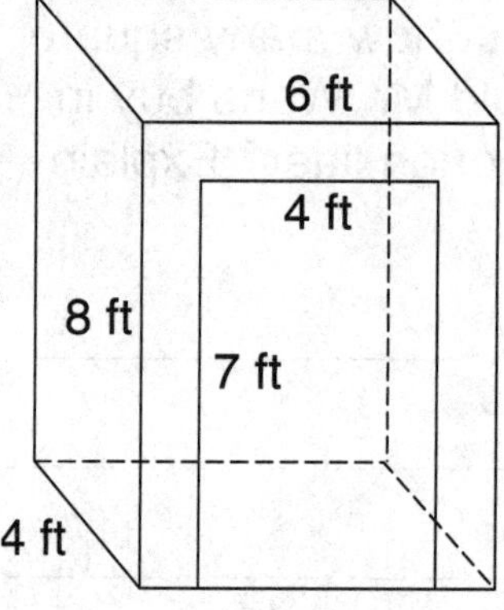

____

Holt Mathematics

Reteach
Surface Area of Prisms and Cylinders

The surface area of a three-dimensional figure is the combined areas of the faces. You can find the surface area of a prism by drawing a **net** of the flattened figure.

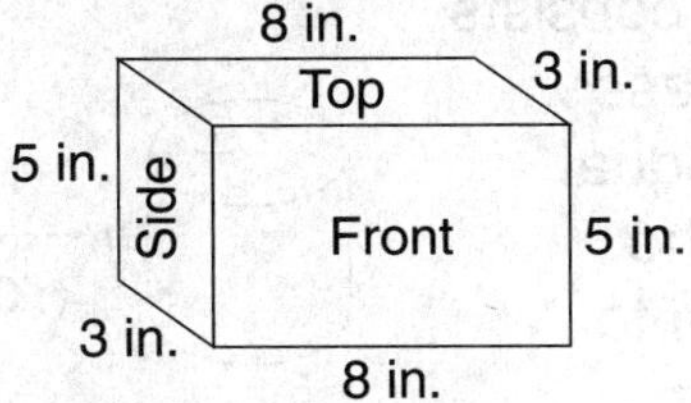

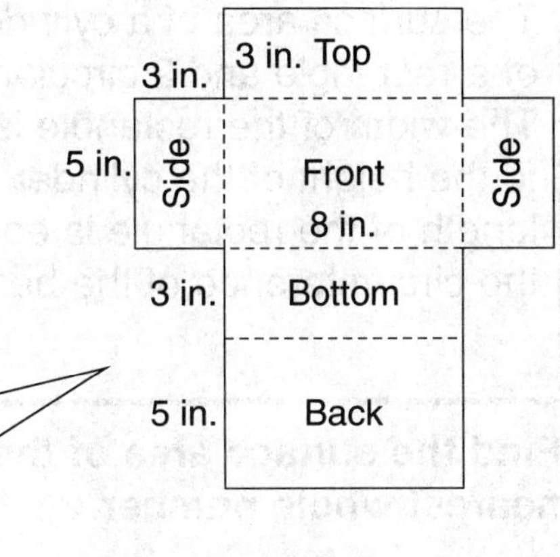

Notice that the front and back faces are the same. The side faces are the same. The top and bottom faces are the same.

Find the surface area of the prism formed by each net.

1. Find the area of the front face: $A =$ ______ • ______ = ______ in^2.

   The area of the front and back faces is 2 • ______ = ______ in^2.

2. Find the area of the side face: $A =$ ______ • ______ = ______ in^2.

   The area of the 2 side faces is 2 • ______ = ______ in^2.

3. Find the area of the top face: $A =$ ______ • ______ = ______ in^2.

   The area of the top and bottom faces is 2 • ______ = ______ in^2.

4. Combine the areas of the faces: ______ + ______ + ______ = ______ in^2.

   The surface area of the prism is ______ in^2.

Find the surface area of the prism formed by each net.

5.

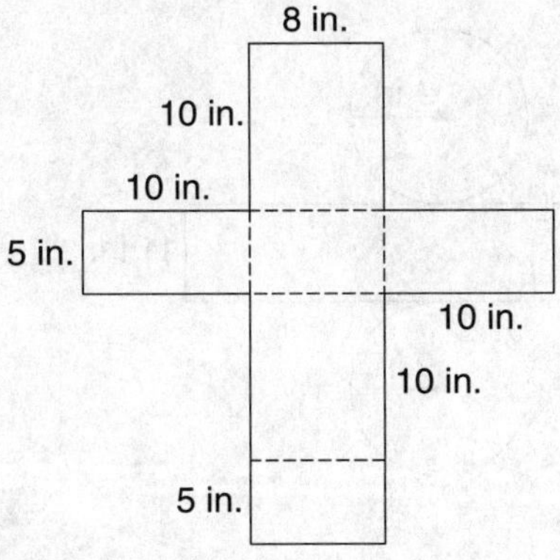

6.

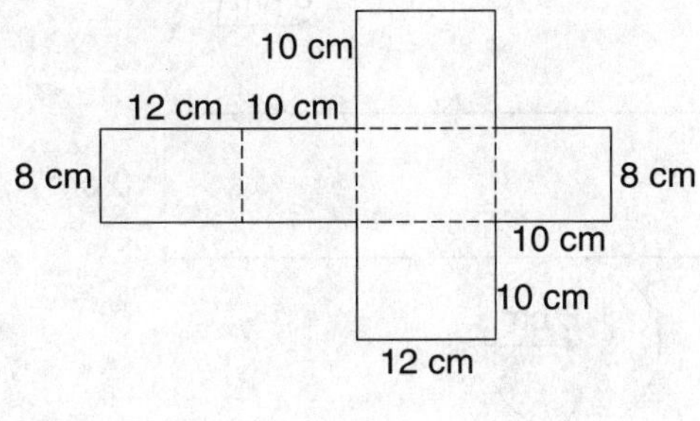

Holt Mathematics

Reteach

Surface Area of Prisms and Cylinders (continued)

The surface area of a cylinder consists of a rectangle and 2 circular bases. The width of the rectangle is equal to the height of the cylinder. The length of the rectangle is equal to the circumference of the bases.

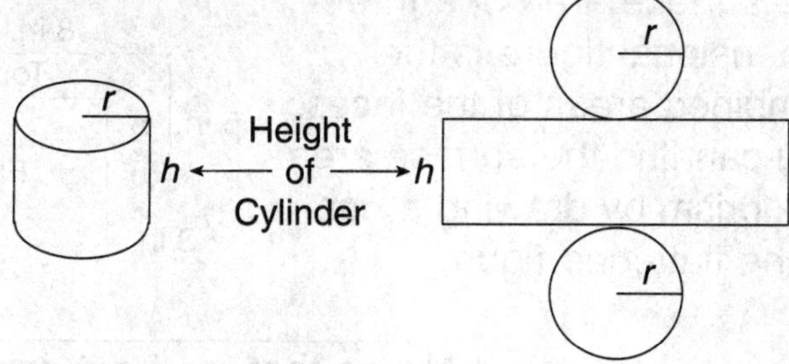

Find the surface area of the cylinder formed by the net to the nearest whole number.

7. Find the area of each circular base.

$$A = \pi r^2 = 3.14 \cdot \underline{\hspace{1cm}} = \underline{\hspace{1cm}} \text{ cm}^2$$

8. Find the area of both circular bases.

$$A = 2 \cdot \underline{\hspace{1cm}} = \underline{\hspace{1cm}} \text{ cm}^2$$

9. Find the area of the rectangle.

$$A = \ell \cdot w$$

$$A = 2\pi r \cdot h$$

$$A = 2 \cdot 3.14 \cdot \underline{\hspace{1cm}} \cdot \underline{\hspace{1cm}}$$

$$A = \underline{\hspace{1cm}} \text{ cm}^2$$

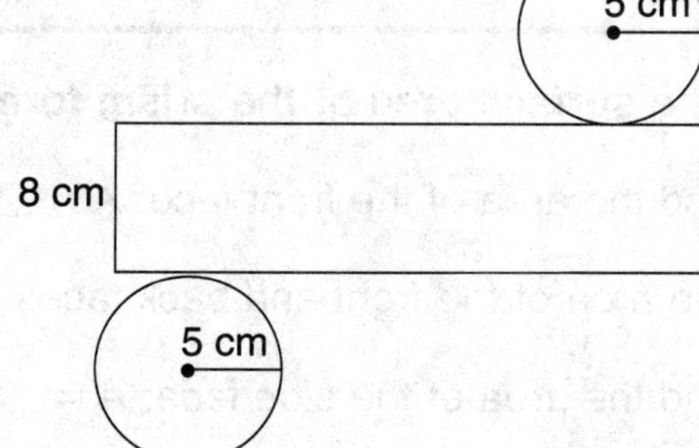

10. Combine the areas of the circular bases and the area of the rectangle to find the surface area, *S*, of the cylinder.

$$S = \underline{\hspace{1cm}} + \underline{\hspace{1cm}} = \underline{\hspace{1cm}} \text{ cm}^2$$

Find the surface area of the cylinder formed by each net. Round to the nearest whole number. Use 3.14 for π.

11.

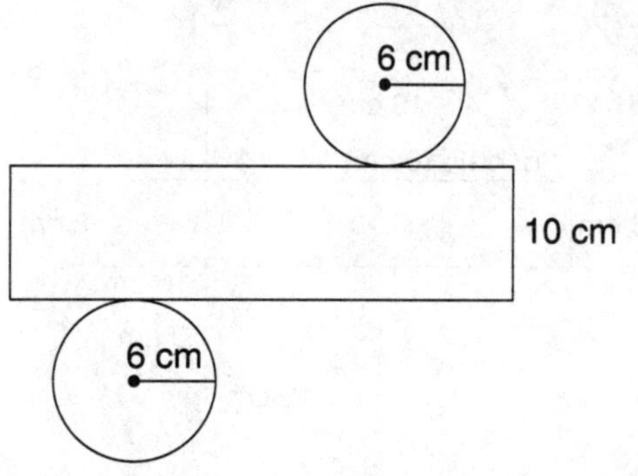

____

12.

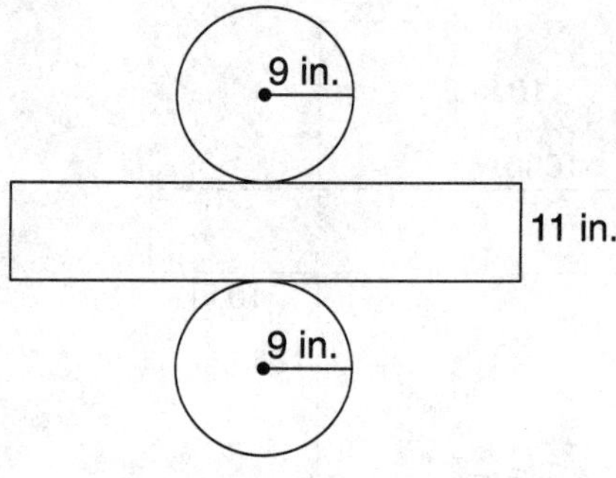

____

Holt Mathematics

<table><tr><td>LESSON
10-4</td><td>

Challenge
Saving Trees
</td></tr></table>

Boxes with different dimensions can have the same volume.
Marcus Manufacturing uses boxes to ship office supplies to stores.
The less cardboard used to make a box, the less the box costs.

**In each set, the boxes have equal volume. Find the missing
dimension, if necessary. Write the surface area below each box.
Then circle the box in each set that uses the least amount of
cardboard.**

1.

surface
area = __________

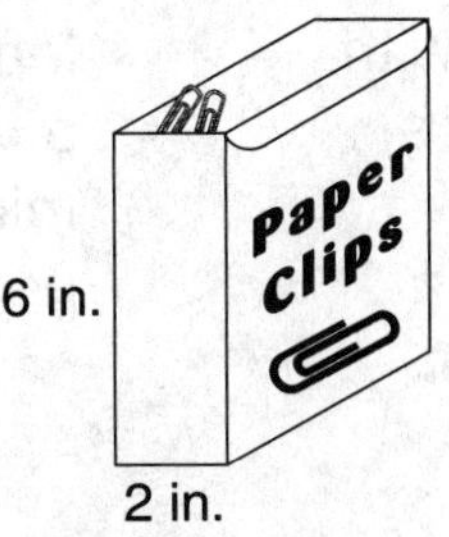

width = __________

surface
area = __________

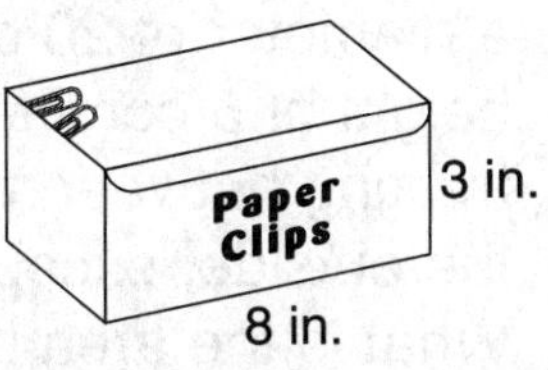

width = __________

surface
area = __________

2.

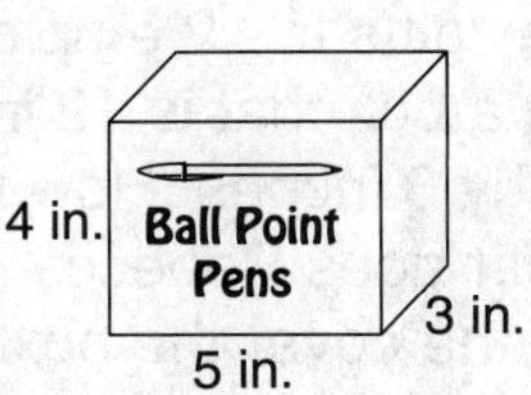

surface
area = __________

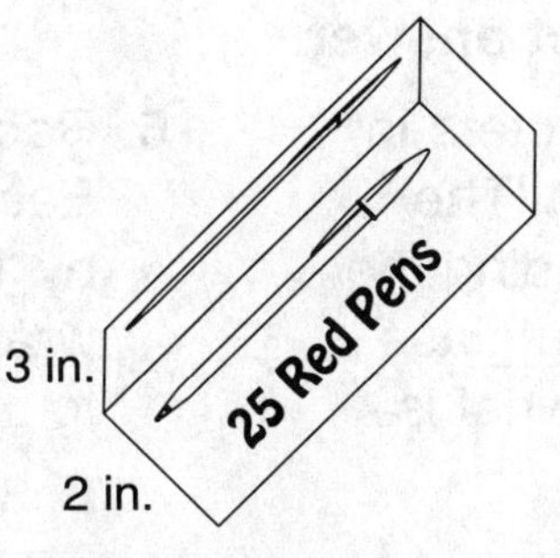

width = __________

surface
area = __________

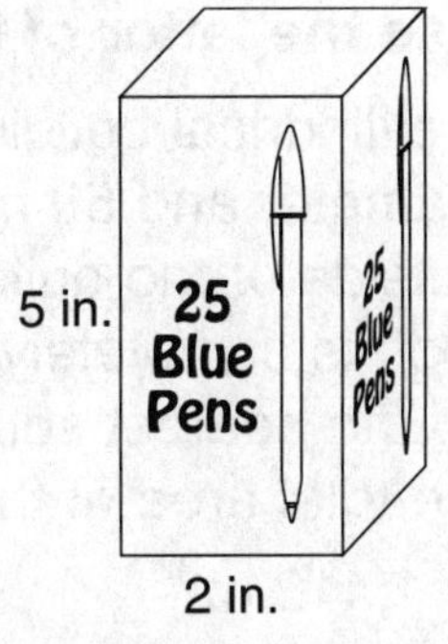

width = __________

surface
area = __________

3.

width = __________
surface
area = __________

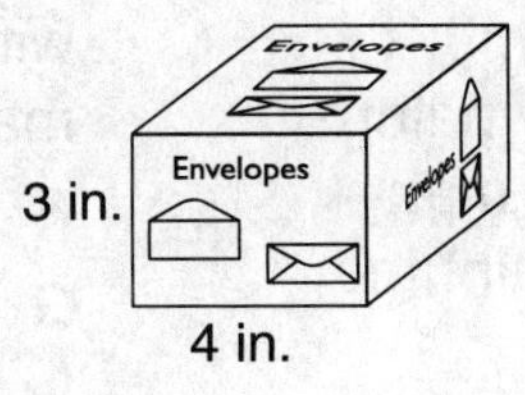

width = __________
surface
area = __________

width = __________
surface
area = __________

Holt Mathematics

LESSON 10-4 **Problem Solving**
Surface Area of Prisms and Cylinders

Write the correct answer.

1. A can of peas is 3 inches in diameter and 4.5 inches tall. What is the area of the label used around the can?

2. How much wrapping paper do you need to completely cover a rectangular box that is 20 inches by 18 inches by 2 inches?

3. Jan puts frosting on a circular cake. The cake has three layers, each with a diameter of 20 centimeters and a height of 5 centimeters. Jan puts frosting between the layers and on the outside, except for the bottom. What is the area that Jan frosts? Round to the nearest square centimeter.

4. A cardboard storage carton has a length of 3 feet, a width of 2 feet, and a volume of 12 ft^3. What is the minimum amount of cardboard needed to make the box?

Choose the letter of the correct answer.

5. A cylindrical building is 30 meters in diameter and 50 meters high. The outside of the building, excluding the roof, is completely covered in glass. To the nearest square foot, what is the total area of the glass?

   A 1,413 m^3 **C** 6,113 m^3

   B 4,710 m^3 **D** 9,420 m^3

6. Rebecca gives gifts to 12 employees. Each gift is in a box that is 12 inches by 10 inches by 3 inches. How much wrapping paper does Rebecca need to completely the cover the boxes?

   F 360 in^2 **H** 4,320 in^2

   G 1,720 in^2 **J** 14,400 in^2

7. A cylinder-shaped sculpture is 24 meters high with a diameter of 6.8 meters. An artist plans to spray-paint the entire surface with silver paint. If one can of spray paint covers 50 square meters, how many cans does the artist need to paint the sculpture?

   A 51 cans   **C** 12 cans

   B 22 cans   **D** 10 cans

8. A rectangular sofa cushion is 36 inches by 30 inches by 20 inches. How many cushions can be covered with 38,400 square inches of material?

   F 6 cushions   **H** 9 cushions

   G 8 cushions   **J** 10 cushions

Holt Mathematics

LESSON 10-4 | Reading Strategies
Analyze Information

If you unfold a three-dimensional figure and lay it flat, you have
made a net.

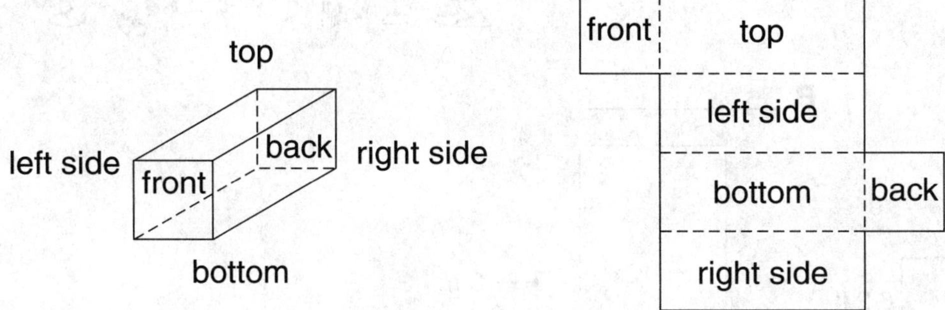

A **net** is a two-dimensional shape that lets you picture all the faces
of a three-dimensional figure. A net helps you see how much
surface a three-dimensional figure covers.

The **surface area** of a three-dimensional figure is the total area of
all its faces.

If you analyze the net of a rectangular prism, you notice there are
six faces. Each face pairs up with another, congruent face:

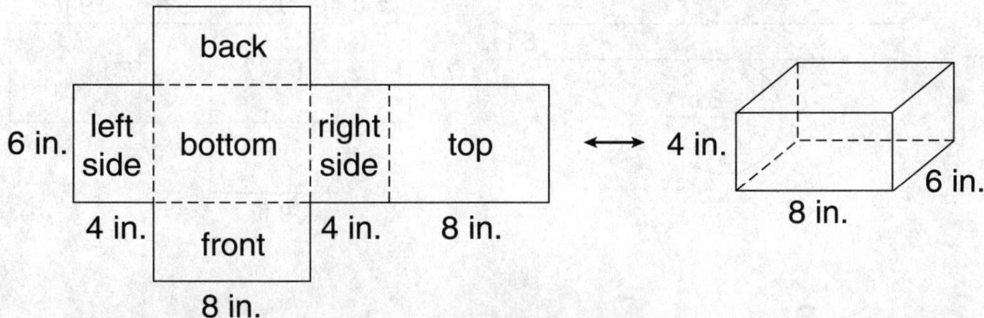

To find the surface area of a rectangular prism, find the sum of the
areas of the six faces (rectangles). *Hint:* $A = \ell w$.

Answer the following to find the surface area of the rectangular prism.

1. What is the area of the bottom rectangle? _______________

2. What is the area of the top rectangle? _______________

3. What is the area of the back rectangle? _______________

4. What is the area of the front rectangle? _______________

5. What is the area of the right rectangle? _______________

6. What is the area of the left rectangle? _______________

7. What is the surface area of the rectangular prism? _______________

Holt Mathematics

Puzzles, Twisters & Teasers

LESSON 10-4

Bee a Math Marvel!

Find the surface area of each figure below. Round to the nearest tenth. Use your answers to solve the riddle.

L _______________

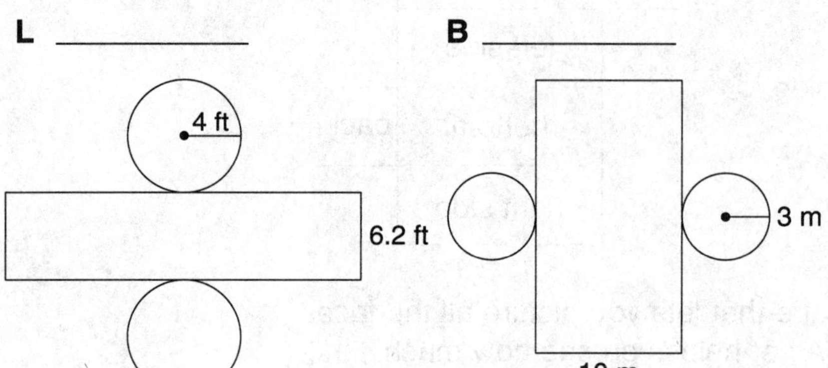

B _______________

A _______________

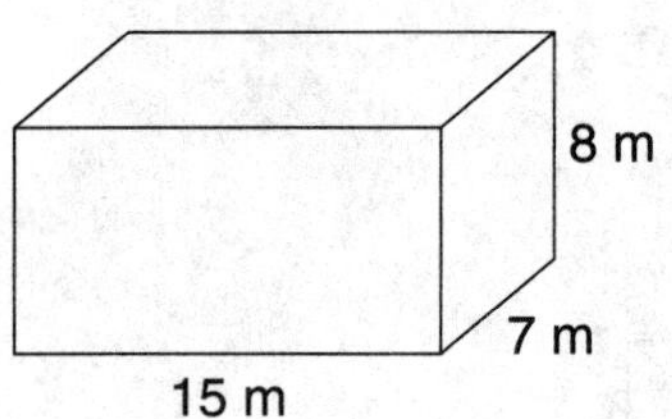

H _______________

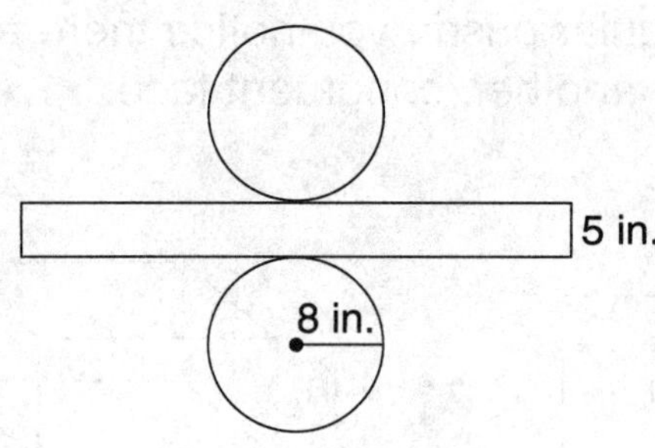

T _______________

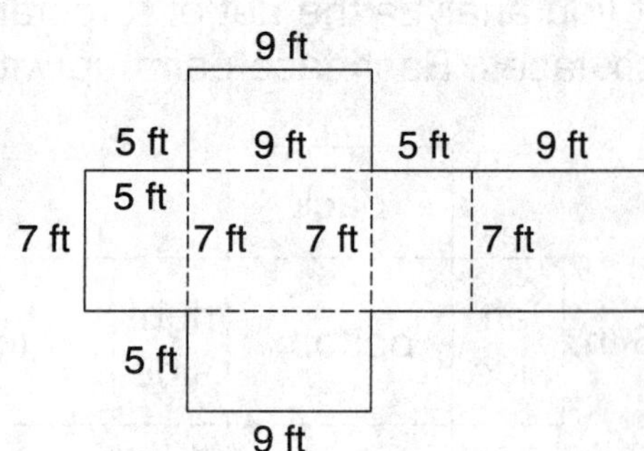

O _______________

S _______________

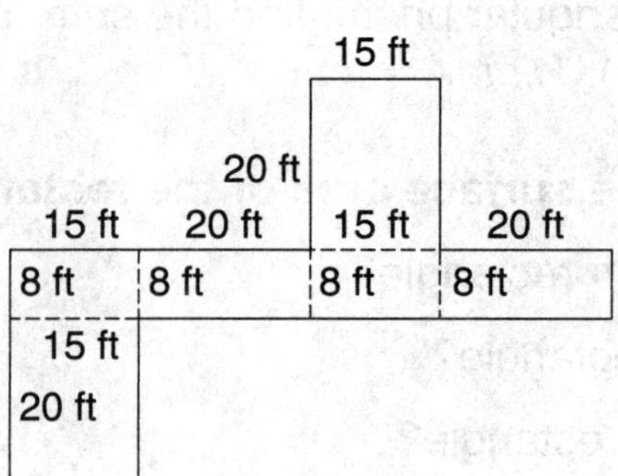

I _______________

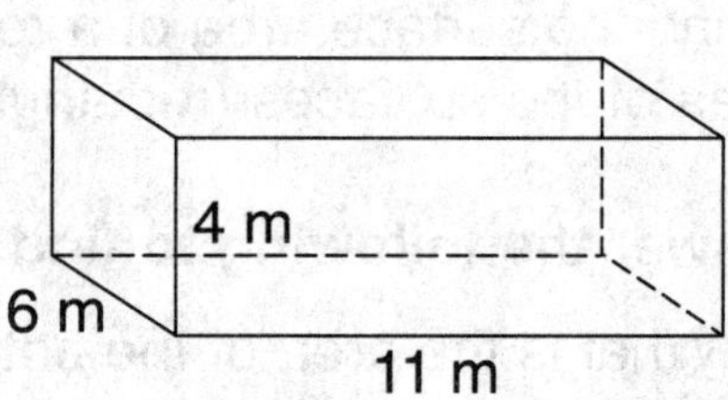

Why couldn't the bee go to the dance?

___ ___ **W** ___ ___ ___ **M** ___ ___ ___ ___ ___ ___ ___

264 286   562 1,160 562   351.7 286 653.1 244.9 562 256.2 256.2

Holt Mathematics

LESSON 10-5 Practice A
Changing Dimensions

Two prisms are similar. The scale factor and the surface area of the smaller prism are given in Column A. Draw a line to the surface area of the larger prism in Column B.

Column A Prisms	Column B Prisms

1. scale factor: 3
surface area of the
smaller prism: 9 cm^2

    A 96 cm^2

2. scale factor: 4
surface area of the
smaller prism: 5 cm^2

    B 90 cm^2

3. scale factor: 5
surface area of the
smaller prism: 4 cm^2

    C 80 cm^2

    D 100 cm^2

4. scale factor: 2
surface area of the
smaller prism: 24 cm^2

    E 81 cm^2

Two prisms are similar. The scale factor and the volume of the smaller prism are given in Column A. Draw a line to the volume of the larger prism in Column B.

Column A Prisms	Column B Prisms

5. scale factor: 3
volume of the smaller
prism: 5 ft^3

    F 125 ft^3

6. scale factor: 4
volume of the smaller
prism: 2 ft^3

    G 128 ft^3

7. scale factor: 5
volume of the smaller
prism: 1 ft^3

    H 135 ft^3

    I 136 ft^3

8. scale factor: 2
volume of the smaller
prism: 17 ft^3

    J 144 ft^3

9. A fish tank holds 6 gallons of water. A larger, similarly
shaped fish tank has a scale factor of 2. How many
times more water does the larger fish tank hold than
the smaller fish tank? _______________

Holt Mathematics

Practice B
Changing Dimensions

Given the scale factor, find the surface area of the similar prism.

1. The scale factor of two similar rectangular prisms is 3. The surface area of the smaller prism is 15 in^2. _______________

2. The scale factor of two similar triangular prisms is 2. The surface area of the smaller prism is 25 cm^2. _______________

3. The scale factor of two similar rectangular prisms is $\frac{1}{2}$. The surface area of the larger prism is 960 ft^2. _______________

4. The scale factor of two similar triangular prisms is 5. The surface area of the smaller prism is 10 m^2. _______________

5. The scale factor of two similar rectangular prisms is $\frac{1}{5}$. The surface area of the larger prism is 625 in^2. _______________

6. The scale factor of two similar pentagonal prisms is 4. The surface area of the smaller prism is 16 cm^2. _______________

Given the scale factor, find the volume of the similar prism.

7. The scale factor of two similar triangular prisms is 3. The volume of the smaller prism is 8 in^3. _______________

8. The scale factor of two similar rectangular prisms is $\frac{1}{2}$. The volume of the larger prism is 648 m^3. _______________

9. The scale factor of two similar triangular prisms is 4. The volume of the smaller prism is 10 cm^3. _______________

10. The scale factor of two similar rectangular prisms is $\frac{1}{4}$. The volume of the larger prism is 1,920 ft^3. _______________

11. The scale factor of two similar triangular prisms is 2. The volume of the smaller prism is 72 yd^3. _______________

12. A small tank weighs 24 pounds when it is full of water. A larger tank that is similar in shape has a scale factor of 3. How much does the larger tank weigh when filled with water? _______________

Holt Mathematics

Practice C
Changing Dimensions

Given the scale factor, find the surface area of the similar prism.

1. The scale factor of two similar rectangular prisms is 5.
   The surface area of the smaller prism is 13 in^2. _______________

2. The scale factor of two similar triangular prisms is $\frac{1}{8}$.
   The surface area of the larger prism is 1,024 cm^2. _______________

3. The scale factor of two similar rectangular prisms is 4.
   The surface area of the smaller prism is 32 ft^2. _______________

4. The scale factor of two similar triangular prisms is $\frac{1}{3}$.
   The surface area of the larger prism is 828 m^2. _______________

5. The scale factor of two similar rectangular prisms is 3.
   The surface area of the smaller prism is 13.5 yd^2. _______________

Given the scale factor, find the volume of the similar prism.

6. The scale factor of two similar triangular prisms is 5.
   The volume of the smaller prism is 12 in^3. _______________

7. The scale factor of two similar rectangular prisms is 3.
   The volume of the smaller prism is 75 m^3. _______________

8. The scale factor of two similar triangular prisms is $\frac{1}{2}$.
   The volume of the larger prism is 976 cm^3. _______________

9. The scale factor of two similar rectangular prisms is 5.
   The volume of the smaller prism is 2.5 ft^3. _______________

10. The scale factor of two similar triangular prisms is $\frac{1}{4}$.
    The volume of the larger prism is 1,120 cm^3. _______________

11. A large fish tank is made of 2,376 square inches of glass.
    A smaller, similarly shaped fish tank has a scale factor of
    $\frac{1}{2}$. Did it take more or less than 1,000 square inches of
    glass to make the smaller fish tank? Explain your answer.

____

____

Holt Mathematics

LESSON
Reteach
10-5 Changing Dimensions

Changing the dimensions of a
three-dimensional figure affects
its surface area and its volume.

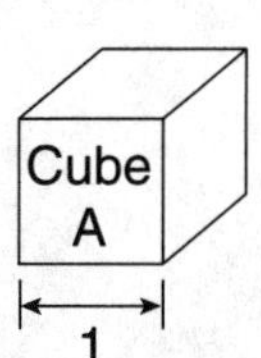

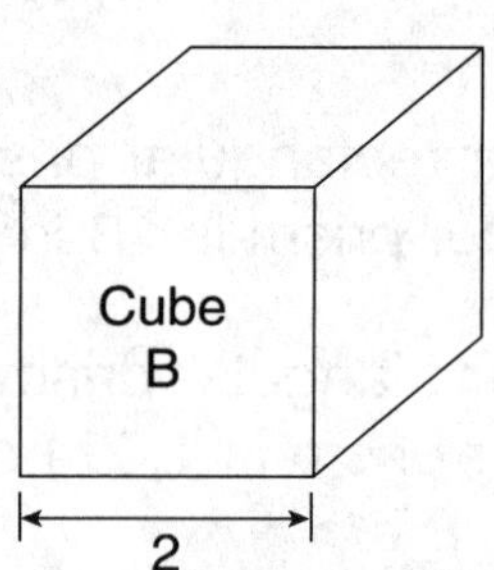

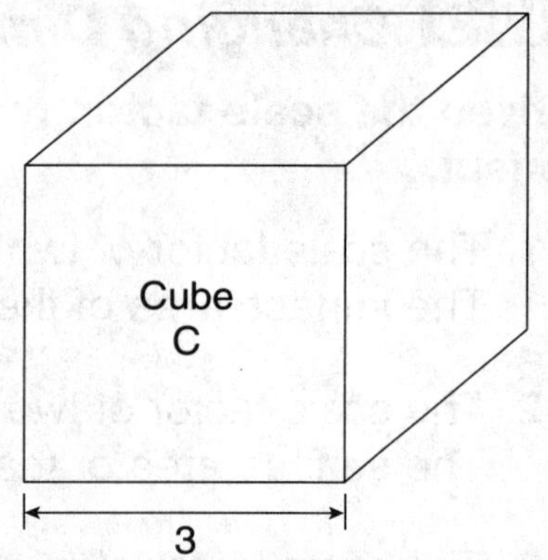

1. Complete the table for the cubes shown above.

Length of Side	Area of One Face	Surface Area	Volume
1	unit^2	units^2	unit^3
2	units^2	units^2	units^3
3	units^2	units^2	units^3

2. The length of one edge of B is _______ times the length of one edge of A.

The area of one face of B is ______ times the area of one face of A.

The surface area of B is ______ times the surface area of A: $2^2 =$ ______.

The volume of B is ______ times the volume of A: $2^3 =$ ______.

3. The length of one edge of C is ______ times the length of one edge of A.

The area of one face of C is ______ times the area of one face of A.

The surface area of C is ______ times the surface area of A: $3^2 =$ ______.

The volume of C is ______ times the volume of A: $3^3 =$ ______.

For any set of similar three-dimensional figures:

• The surface area of the larger figure equals the surface area of the
smaller figure times the scale factor squared.

• The volume of the larger figure equals the volume of the smaller figure
times the scale factor cubed.

The scale factor of two similar prisms is 4.

4. The surface area of the smaller prism is 9 in^2. Multiply by ______, or ______.

The surface area of the larger prism is ______ in^2.

Holt Mathematics

 Challenge
 Eggs-actly

The volumes of two similar-shaped figures
are proportional to the cube of their lengths.

Suppose an extra-large egg is 2.5 inches long,
compared to a smaller egg 1.25 inches long.
What is the ratio of their volumes?

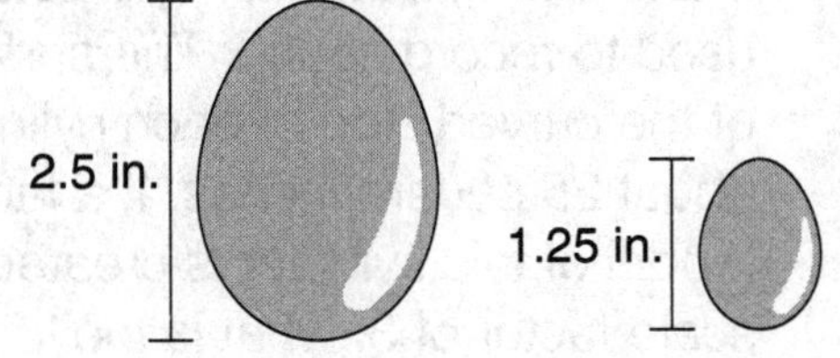

Cube the ratio of their lengths.

$$\left(\frac{2.5}{1.25}\right)^3 = \left(\frac{2}{1}\right)^3 = \frac{8}{1}$$

The ratio of their volumes is 8:1.

So, the larger egg has 8 times as much volume, 8 times as much
weight, and 8 times as much food.

Solve. Show your work.

1. One egg is 2 inches long. Another is 2.75 inches long. How
   many times greater is the volume of the larger egg?

   ____

2. One egg is 2.25 inches long. Another is 1.5 inches long. How
   many times greater is the weight of the larger egg?

   ____

3. The ratio of the lengths of two eggs is 5 to 2. The larger egg
   contains how many times the food of the smaller egg?

   ____

4. An egg is 2 inches long. What is the approximate length of an
   egg with twice the volume?

   ____

5. An egg is 3 inches long. What is the approximate length of an
   egg with half the weight?

   ____

6. An egg is 1.75 inches long. What is the approximate length of
   an egg with twice the food?

   ____

Holt Mathematics

Problem Solving
Changing Dimensions

Write the correct answer.

1. In the late 1800s, wax cylinders were used to record sound. The surface area of the curved side of each cylinder was about 25 square inches. If a larger model of this cylinder is created using a scale factor of 3, what is the lateral surface area of the model?

2. A 5-foot wide, 100-foot tall cylindrical water tower was built in St. Louis in the early 1800s. An architecture student wants to build a model using a scale factor of $\frac{1}{6}$. What will be the volume of the model to the nearest cubic foot?

3. A cone-shaped plastic cup holds 24 ounces of water. A smaller cup has a scale factor of $\frac{1}{2}$. How much water does the smaller cup hold?

4. In the game of Ring Taw, players use a shooting marble that has a surface area of 1.77 square inches. What is the surface area of a large ball if the scale factor is 5?

Choose the letter of the correct answer.

5. The volume of a rectangular prism is 48 cubic centimeters. The volume of a similar rectangular prism is 6 cubic centimeters What is the scale factor for the rectangles?

   A 8 **C** 4

   B 6 **D** 2

6. A cooking pot used in the cafeteria weighs 64 pounds when it is filled with soup. How much would a similar pot with a scale factor of $\frac{1}{2}$ weigh when filled with the same soup?

   F 4 lb **H** 16 lb

   G 8 lb **J** 32 lb

7. For his science project, Marty is building a model of Pluto, which has a surface area of about 6,376,000 square miles. He plans to cover his model with red foil. If he uses a scale factor of 5,000, how much red foil will he need to the nearest hundredth of a square mile?

   A 1.26 mi^2 **C** 0.26 mi^2

   B 0.026 mi^2 **D** 0.126 mi^2

8. The scale factor of two similar triangular prisms is 5. What is the possible surface area of both prisms?

   F 2,000 cm^2 and 80 cm^2

   G 125 cm^2 and 25 cm^2

   H 5,000 cm^2 and 40 cm^2

   J 10,000 cm^2 and 60 cm^2

Holt Mathematics

Reading Strategies
LESSON 10-5 *Organize Information*

Each face on this cube measures 1 unit by 1 unit. The area of one face is 1 square unit.

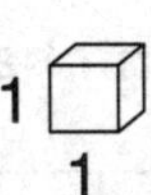

Adding 6 faces = 6 square units. The surface area is 6 units^2.

Each face on this cube measures 2 units by 2 units.

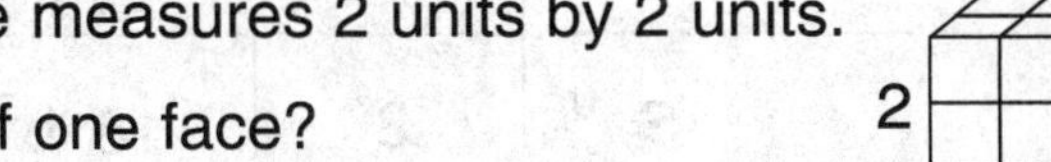

1. What is the area of one face? ______________________

2. What is the total area of all six faces? ______________________

You can use a table to organize the information about the two cubes.

Dimensions of Each Face	Area of 1 Face	Total Surface Area
1 unit by 1 unit	1 unit^2	$6 \cdot 1 = 6$ units^2
2 units by 2 units	4 units^2	$6 \cdot 4 = 24$ units^2

This cube would be the next cube in the list.

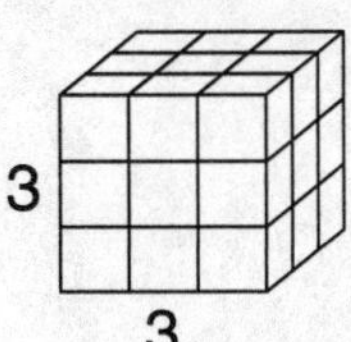

Use this cube to answer the following questions.

3. What are the dimensions of each face on the cube? ______________

4. What is the area of one face of the cube?

____

5. How many faces does the cube have? ______________

6. What is the surface area for the cube?

____

7. If the list above continued, what would be the dimensions of each face on the next cube? ______________

Holt Mathematics

<table>
<tr><td>**LESSON 10-5**</td></tr>
</table>

Puzzles, Twisters & Teasers

A New Dimension!

Complete both charts. Use your answers to solve the riddle.

scale factor	smaller surface area	larger surface area
4	35 in^2	**T**
6	24 m^2	**S**
8	40 ft^2	**U**
9	**H**	1.602 m^2

scale factor	smaller volume	larger volume
2	29 ft^3	**C**
3	18 lb^3	**E**
5	8 cm^3	**R**
4	17 in^3	**O**
4	32 lb^3	**A**
2	**K**	600 in^3

Why did the hairdresser win the race?

___ ___ ___   ___ ___ ___ ___   ___
864   20   486   560 1,088 1,088 75   2,048

___ ___ ___ ___ ___ ___ ___ ___.
864   20 1,088 1,000 560 232 2,560 560

44

Holt Mathematics

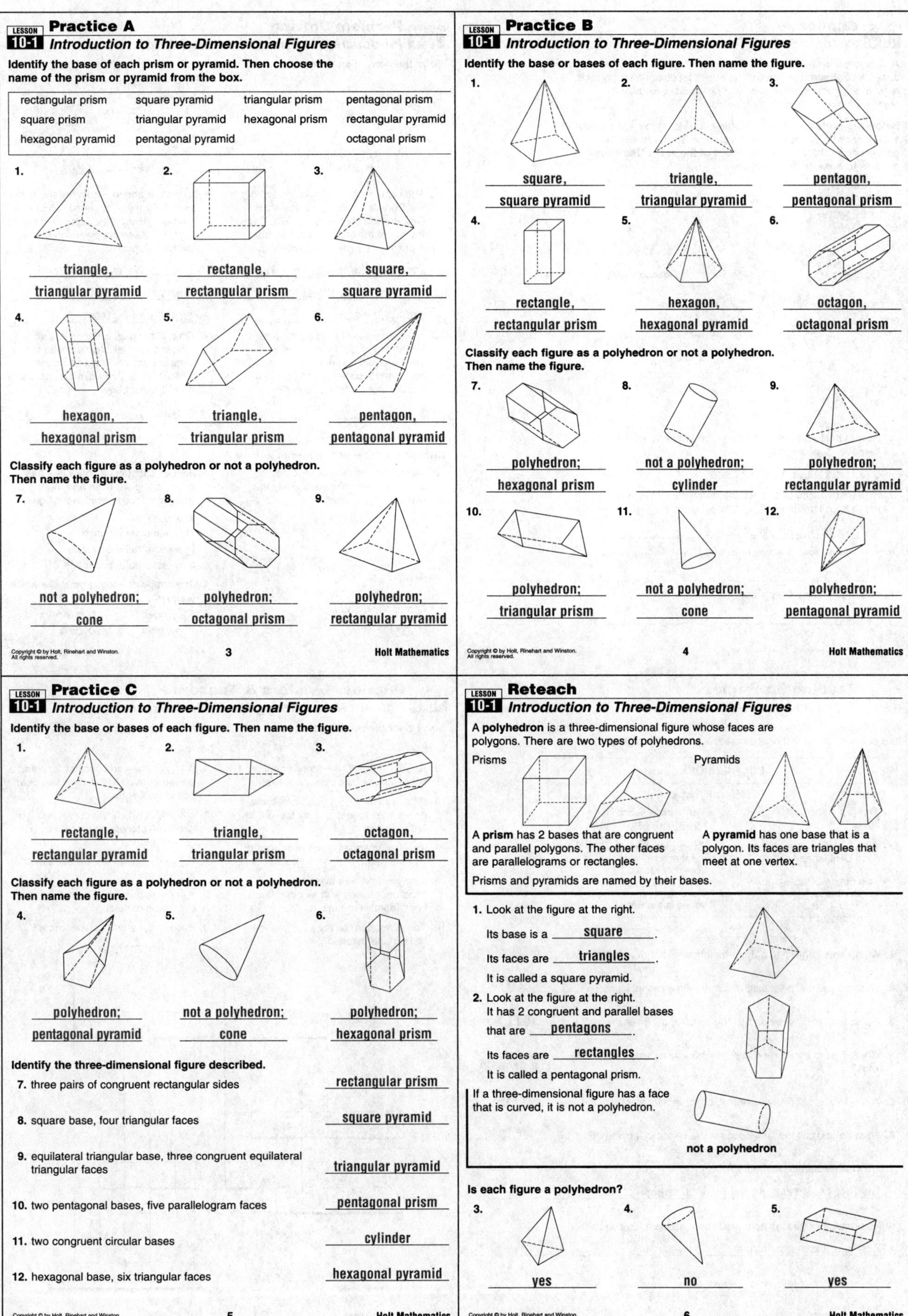

Practice A
10-1 Introduction to Three-Dimensional Figures

Identify the base of each prism or pyramid. Then choose the name of the prism or pyramid from the box.

rectangular prism	square pyramid	triangular prism	pentagonal prism
square prism	triangular pyramid	hexagonal prism	rectangular pyramid
hexagonal pyramid	pentagonal pyramid		octagonal prism

1.
triangle,
triangular pyramid

2.
rectangle,
rectangular prism

3.
square,
square pyramid

4.
hexagon,
hexagonal prism

5.
triangle,
triangular prism

6.
pentagon,
pentagonal pyramid

Classify each figure as a polyhedron or not a polyhedron. Then name the figure.

7.
not a polyhedron;
cone

8.
polyhedron;
octagonal prism

9.
polyhedron;
rectangular pyramid

3
Holt Mathematics

Practice B
10-1 Introduction to Three-Dimensional Figures

Identify the base or bases of each figure. Then name the figure.

1.
square,
square pyramid

2.
triangle,
triangular pyramid

3.
pentagon,
pentagonal prism

4.
rectangle,
rectangular prism

5.
hexagon,
hexagonal pyramid

6.
octagon,
octagonal prism

Classify each figure as a polyhedron or not a polyhedron. Then name the figure.

7.
polyhedron;
hexagonal prism

8.
not a polyhedron;
cylinder

9.
polyhedron;
rectangular pyramid

10.
polyhedron;
triangular prism

11.
not a polyhedron;
cone

12.
polyhedron;
pentagonal pyramid

4
Holt Mathematics

Practice C
10-1 Introduction to Three-Dimensional Figures

Identify the base or bases of each figure. Then name the figure.

1.
rectangle,
rectangular pyramid

2.
triangle,
triangular prism

3.
octagon,
octagonal prism

Classify each figure as a polyhedron or not a polyhedron. Then name the figure.

4.
polyhedron;
pentagonal pyramid

5.
not a polyhedron;
cone

6.
polyhedron;
hexagonal prism

Identify the three-dimensional figure described.

7. three pairs of congruent rectangular sides
rectangular prism

8. square base, four triangular faces
square pyramid

9. equilateral triangular base, three congruent equilateral triangular faces
triangular pyramid

10. two pentagonal bases, five parallelogram faces
pentagonal prism

11. two congruent circular bases
cylinder

12. hexagonal base, six triangular faces
hexagonal pyramid

5
Holt Mathematics

Reteach
10-1 Introduction to Three-Dimensional Figures

A **polyhedron** is a three-dimensional figure whose faces are polygons. There are two types of polyhedrons.

Prisms

Pyramids

A **prism** has 2 bases that are congruent and parallel polygons. The other faces are parallelograms or rectangles.

A **pyramid** has one base that is a polygon. Its faces are triangles that meet at one vertex.

Prisms and pyramids are named by their bases.

1. Look at the figure at the right.

Its base is a ___square___.

Its faces are ___triangles___.

It is called a square pyramid.

2. Look at the figure at the right.
It has 2 congruent and parallel bases that are ___pentagons___.

Its faces are ___rectangles___.

It is called a pentagonal prism.

If a three-dimensional figure has a face that is curved, it is not a polyhedron.

not a polyhedron

Is each figure a polyhedron?

3.
yes

4.
no

5.
yes

6
Holt Mathematics

Holt Mathematics

A three-dimensional figure in which all the faces are polygons is called a *polyhedron*. A polyhedron whose faces are all congruent regular polygons is called a *regular polyhedron*. Regular polyhedrons are called *Platonic solids*.

Below are patterns for four Platonic solids. Copy each pattern. You may enlarge the pattern if you like. Then cut out the pattern and fold it to build a model of the solid. Use your models to answer the questions.

Hexahedron

Tetrahedron

Octahedron

Dodecahedron

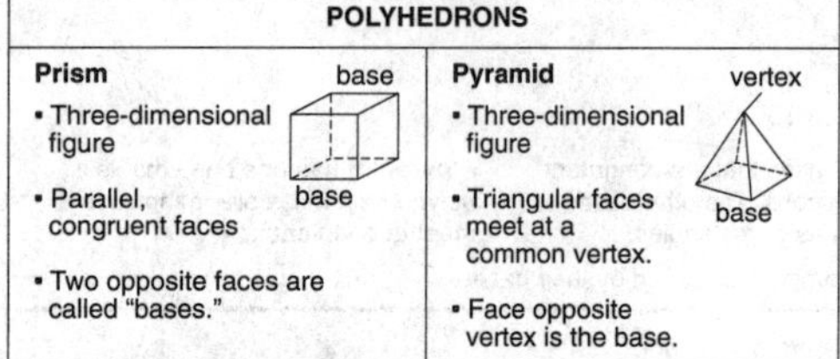

1. How many faces does a hexahedron have? What is another name for a hexahedron?

   6 faces; square prism or cube

2. How many faces does a tetrahedron have? What is another name for a tetrahedron?

   4 faces; triangular pyramid

3. How many faces does an octahedron have?

   8 faces

4. How many faces does a dodecahedron have?

   12 faces

7   **Holt Mathematics**

Write the correct answer.

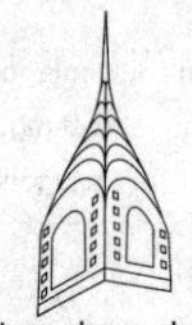
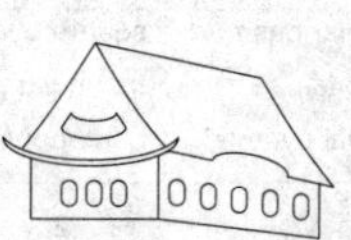

1. The picture above shows the top of the Chrysler Building in New York City. It was completed in 1930. Does the top of the tower most resemble a prism or a pyramid? Explain.

   pyramid; The top of the tower
   is a vertex, and the sides
   are triangular.

2. The picture above shows the rooftop of Himeji Castle, completed in 1614 in Donjon, Himeji City, Japan. Do the rooftops resemble pyramids or prisms? Explain.

   a prism; It has triangular
   bases and parallelograms
   form the other faces.

3. An architect designed a structure for the top of a building. The structure has a vertex, one circular base, and a curved surface. What three-dimensional figure is it?

   a cone

4. On a farm, grain is stored in a silo. This is a very tall structure with a circular base and top and a curved surface. What three-dimensional figure does it resemble?

   a cylinder

Choose the letter of the correct answer.

5. James put two blocks together to build the figure shown. Identify the two figures he used.

   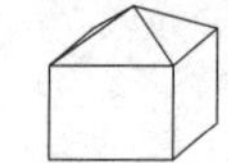

   A two pyramids
   (B) a pyramid and a prism
   C a pyramid and a cone
   D two prisms

6. Jaime constructs a figure that has one rectangular base and three triangular faces. What is the figure?

   F a cone
   G a triangular pyramid
   (H) a rectangular pyramid
   J a rectangular prism

7. The shape of a log is most like which figure?
   (A) cylinder   C prism
   B pyramid   D cone

8   **Holt Mathematics**

A **polyhedron** is a three-dimensional figure with flat sides called **faces**. In a polyhedron, all of the faces are made only of straight sides.

POLYHEDRONS	
Prism base • Three-dimensional figure • Parallel, congruent faces base • Two opposite faces are called "bases."	**Pyramid** vertex • Three-dimensional figure • Triangular faces meet at a common vertex. base • Face opposite vertex is the base.

1. What name is given to a three-dimensional figure with flat sides?   **polyhedron**

2. What do you call a flat side of a three-dimensional figure?   **a face**

3. Which type of polyhedron has two parallel, congruent faces?   **a prism**

4. Which type of polyhedron has a vertex and a face at opposite ends?   **a pyramid**

5. What do you call the two parallel, congruent faces of a prism?   **bases**

6. What do you call the face opposite the vertex in a pyramid?   **the base**

7. Why isn't a cone a polyhedron?

   The faces of a cone are not made of straight sides.

8. Name another three-dimensional figure that is not a polyhedron.

   Possible answer: cylinder

9   **Holt Mathematics**

Solve the crossword puzzle.

Across

3. A ______ is a three-dimensional figure whose faces are all polygons.

6. A hexagonal ______ has one base that is a hexagon, and six faces that are triangles.

7. a line segment at which two faces of a prism meet

8. a polyhedron with two parallel congruent bases and all other faces being parallelograms

10. The face opposite the ______ is the base of the pyramid.

Down

1. A ______ has one circular face and a curved surface that comes to a point called the vertex.

2. a figure in which two congruent circular faces are connected by a curved surface

4. A ______ prism has six faces that are rectangles.

5. A prism is named for its two congruent parallel ______.

9. Each surface of a polyhedron is called a ______.

10   **Holt Mathematics**

Holt Mathematics

Practice A
Volume of Prisms and Cylinders

Find how many cubes each prism holds. Then give the prism's volume.

1. 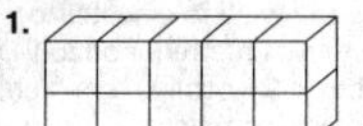

__10 cubes,__
__10 cubic units__

2. 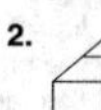

__12 cubes,__
__12 cubic units__

3. 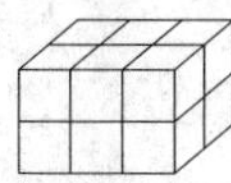

__36 cubes,__
__36 cubic units__

Find the volume of each figure. Choose the letter for the best answer.

4. 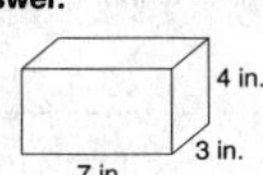

A 14 in^3 C 28 in^3
B 21 in^3 (D) 84 in^3

5. 

F 40 cm^3 (H) 180 cm^3
G 72 cm^3 J 360 cm^3

6. 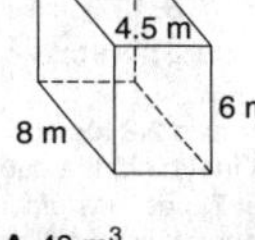

A 48 m^3 (C) 216 m^3
B 162 m^3 D Not Here

7. 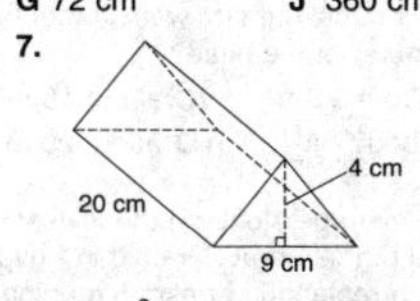

F 60 cm^3 H 720 cm^3
(G) 360 cm^3 J 810 cm^3

8. A juice can is shaped like a cylinder. It is 12 centimeters wide and 16 centimeters tall. Find its volume to the nearest whole number. Use 3.14 for π.

__1,809 cm^3__

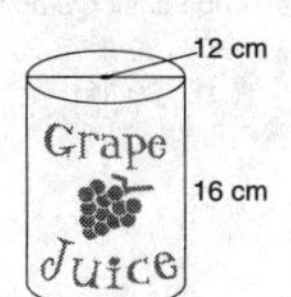

11   **Holt Mathematics**

Practice B
Volume of Prisms and Cylinders

Find how many cubes each prism holds. Then give the prism's volume.

1. 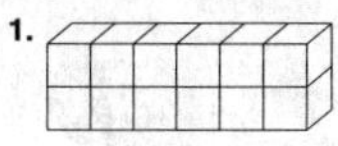

__12 cubes,__
__12 cubic units__

2. 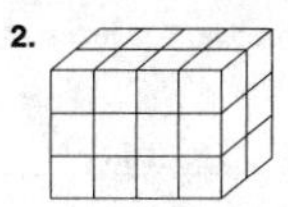

__24 cubes,__
__24 cubic units__

3.

__20 cubes,__
__20 cubic units__

Find the volume of each figure.

4. 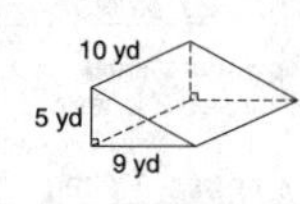

__180 in^3__

5. 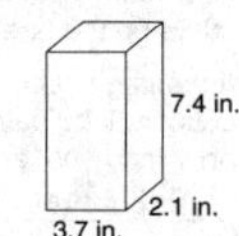

__225 yd^3__

6.

__57.498 in^3__

7.

__729 in^3__

8.

__357.5 cm^3__

9.

__1,053 m^3__

10. A travel mug is shaped like a cylinder. It is 9 centimeters wide and 15 centimeters tall. Find its volume to the nearest tenth. Use 3.14 for π.

__953.8 cm^3__

12   **Holt Mathematics**

Practice C
Volume of Prisms and Cylinders

Find how many cubes each prism holds. Then give the prism's volume.

1. 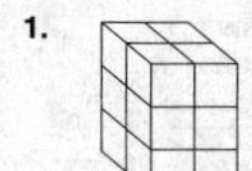

__12 cubes,__
__12 cubic units__

2. 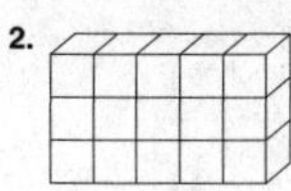

__15 cubes,__
__15 cubic units__

3. 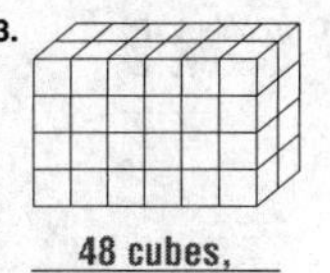

__48 cubes,__
__48 cubic units__

Find the volume of each figure.

4. 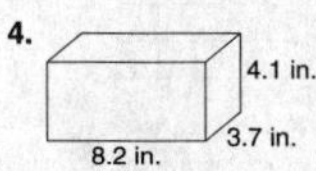

__124.4 in^3__

5. 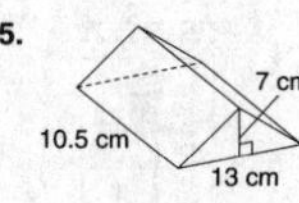

__477.8 cm^3__

6. 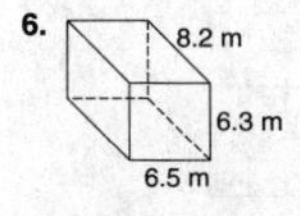

__335.79 m^3__

7. 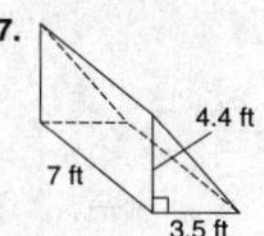

__53.9 ft^3__

8. 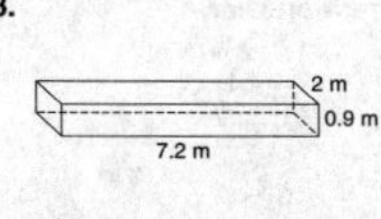

__12.96 m^3__

9.

__20.23 cm^3__

10. A vase is in the shape of a cylinder with a diameter of 5 inches and a height of 9 inches. What is the volume of the vase?   __176.6 in^3__

11. A glass is shaped like a cylinder. The glass has a volume of 602.88 centimeters and a radius of 4 centimeters. What is the height?   __12 cm__

12. The base of a triangular prism is a right triangle with hypotenuse 13 inches long and one leg 12 inches long. The height of the prism is 8 inches. What is the volume of the prism?   __240 in^3__

13   **Holt Mathematics**

Reteach
Volume of Prisms and Cylinders

The **volume** of a three-dimensional figure is the amount of space it takes up. Volume is measured in cubic units.

Find the volume of the prism.

1. Think of the prism as layers of cubes.
   There are __6__ cubes in the bottom layer.
2. There are __3__ layers of cubes.
3. Multiply the number of cubes in the bottom layer by the number of layers.
   The volume of the prism is __18__ cubic units.

The volume of a prism or a cylinder is the area of its base times its height.

volume = base • height, or $V = B \cdot h$

Find the volume of the prism.

4. What is the shape of the base? __triangle__

5. The area of the base is $B = \frac{1}{2}bh$
   $B = \frac{1}{2} \cdot$ __6__ $\cdot$ __3__ $=$ __9__ in^2

6. The height of the prism is __5__ in.

7. $V = B \cdot h =$ __9__ $\cdot$ __5__ $=$ __45__ in^3

Find the volume of the cylinder to the nearest whole number.

8. What is the shape of the base? __circle__

9. The area of the base is $A = \pi r^2$.
   $A = 3.14 \cdot$ __4__ $^2 =$ __50__ cm^2

10. The height of the cylinder is __3__ cm.

11. $V = B \cdot h =$ __50__ $\cdot$ __3__ $=$ __150__ cm^3

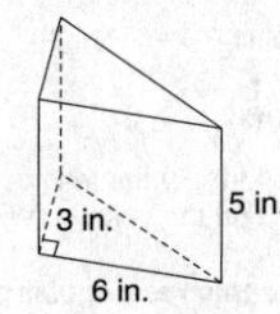

14   **Holt Mathematics**

Challenge
Painted Faces

A cube has six sides, or faces.
Each of the six faces is a square. 

This cube measures 2 units on each edge. The faces
of the cube are painted. 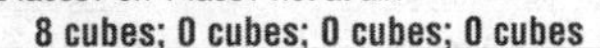

1. How many small cubes
   make up the large cube? __________ **8 cubes**

2. How many small cubes are painted on 3 faces?
   on 2 faces? on 1 face? not at all?
   8 cubes; 0 cubes; 0 cubes; 0 cubes

This cube measures 3 units on each edge. 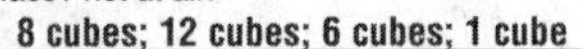

3. How many small cubes
   make up the large cube? __________ **27 cubes**

4. If the large cube is painted, how many small
   cubes will be painted on 3 faces? on 2 faces?
   on 1 face? not at all?
   8 cubes; 12 cubes; 6 cubes; 1 cube

This cube measures 4 units on each edge. 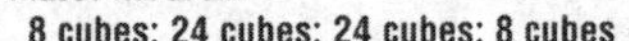

5. How many small cubes
   make up the large cube? __________ **64 cubes**

6. If the large cube is painted, how many small
   cubes will be painted on 3 faces? on 2 faces?
   on 1 face? not at all?
   8 cubes; 24 cubes; 24 cubes; 8 cubes

7. Complete the table below by entering the number
   of small cubes that will be painted on the given
   number of faces.

Size of Large Cube	Painted on 3 Faces	Painted on 2 Faces	Painted on 1 Face	Painted on 0 Faces
2 × 2 × 2	8	0	0	0
3 × 3 × 3	8	12	6	1
4 × 4 × 4	8	24	24	8
5 × 5 × 5	8	36	54	27

8. Look at your table in Exercise 7. What is the rule
   for the number of cubes painted on 3 faces? __________ **always 8**

15
Holt Mathematics

Problem Solving
Volume of Prisms and Cylinders

Write the correct answer.

1. The eight Corinthian columns at the
   National Building Museum in
   Washington, DC, are each 75 feet
   high and 8 feet in diameter. What is
   the volume of each column?
   3,768 ft^3

2. A cubic centimeter holds 1 milliliter of
   liquid. How many liters of water to
   the nearest tenth are required to fill a
   fish tank that is 24 centimeters high,
   28 centimeters long, and
   36 centimeters wide?
   24.2 L

3. There are 231 cubic inches in a
   gallon. A large juice can has a
   diameter of 6 inches and a height of
   10 inches. How many gallons of juice
   does the can hold? Round your
   answer to the nearest tenth.
   1.2 gal

4. A small gift box that holds a ring is
   shaped like a cube. The box
   measures 1.4 inches on each side.
   What is the volume of the gift box?
   Round your answer to the nearest
   tenth.
   2.7 in^3

Choose the letter of the correct answer.

5. The Leaning Tower of Pisa in Italy
   appears to be cylindrical in shape.
   It's height is about 56 meters. If the
   volume of the tower is about
   9,891 cubic meters, what is the
   diameter of the base?
   - A about 3.5 m
   - Ⓒ about 15 m
   - B about 7 m
   - D about 20 m

6. A bricklayer is building a brick
   rectangular post to anchor a mailbox.
   The post is 3 feet tall, 2 feet deep, and
   2 feet wide. Each brick is 3 inches by
   6 inches by 3 inches. How many
   bricks does he need?
   - F 12 bricks
   - H 197 bricks
   - G 54 bricks
   - Ⓙ 384 bricks

7. The average stone on the lowest
   level of the Great Pyramid in Egypt
   was a rectangular prism 5 feet long
   by 5 feet high by 6 feet deep and
   weighed 15 tons. What was the
   volume of the average casing stone?
   - A 1,800 ft^3
   - Ⓒ 150 ft^3
   - B 1,800 ft^2
   - D 150 ft^2

8. A cylindrical barrel is 2.8 feet in
   diameter and 8 feet high. If a cubic
   foot holds about 7.5 gallons of liquid,
   how many gallons of water will this
   barrel hold?
   - F about 1,477 gal
   - Ⓖ about 369 gal
   - H about 150 gal
   - J about 470 gal

16
Holt Mathematics

Reading Strategies
Use a Visual Aid

Think of the **volume** of a solid figure as the number of cubic units
inside the figure.

one cubic unit

Each cube represents one cubic unit.

1. What is the height of the prism in the figure? **3 units**

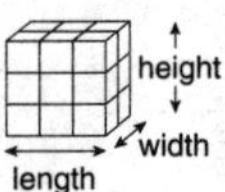

2. What is the width of the prism? **2 units**
3. What is the length of this prism? **3 units**

Multiply the length, width, and height of a prism to find its volume in
cubic units.

Volume = length • width • height

$$V \quad = \quad \ell \quad • \quad w \quad • \quad h$$

4. Multiply the length, width, and height of the prism above to
   find the volume. What is the volume? **18 cubic units**

Use this rectangular prism to complete Exercises 5–7.

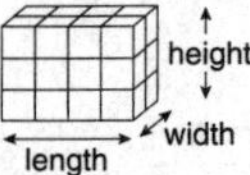

5. How long is the prism? How wide is the prism? **4 units; 2 units**
6. What is the height of the prism? **3 units**
7. Use the formula $V = \ell • w • h$ to find the volume of this prism.
   4 • 2 • 3 = 24 cubic units

17
Holt Mathematics

Puzzles, Twisters & Teasers
Crack the Code!

**Find how many cubes each prism holds. Circle the letter of
the correct answer.**

1. 16 cubes Ⓟ
   12 cubes L
   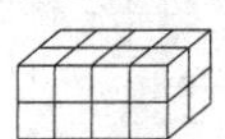

2. 12 cubes Ⓨ
   16 cubes P
   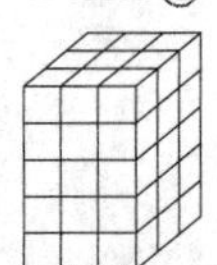

3. 12 cubes Y
   24 cubes Ⓤ
   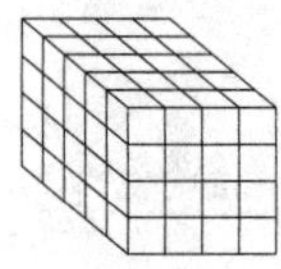

4. 54 cubes A
   45 cubes Ⓞ

5. 80 cubes Ⓔ
   60 cubes I

6. 20 cubes Ⓒ
   16 cubes S
   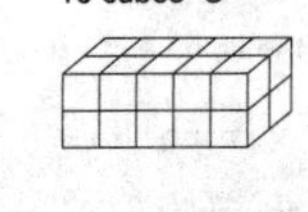

**Find the volume of each figure to the nearest whole number.
Circle the letter of the correct answer.**

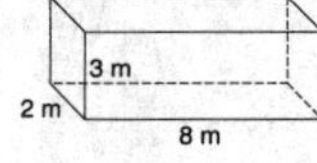

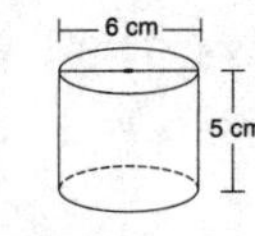

7. 46 m^3 N
   48 m^3 Ⓜ

8. 38 in^3 Ⓡ
   151 in^3 H

9. 47 cm^3 P
   141 cm^3 Ⓚ

Write the letters above the answers to solve the riddle.

What did the egg say to the clown?

Y	O	U	C	R	A	C	K
12	45	24	20	38		20	141

M	E	U	P
48	80	24	16

18
Holt Mathematics

Practice A
Volume of Pyramids and Cones

1. Write the formula for the volume of a pyramid. $V = \frac{1}{3}Bh$

2. Write the formula for the volume of a cone. $V = \frac{1}{3}\pi r^2 h$

Find the volume of each pyramid or cone to the nearest whole number. Use 3.14 for π. Cross out each number in the box that matches a volume.

2,826	34	8	500	360	36
471	6	33	7	38	540

3. 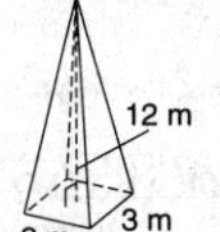
36 m³

4. 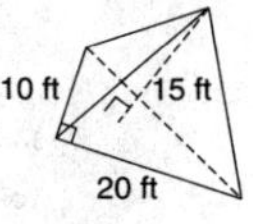
500 ft³

5. 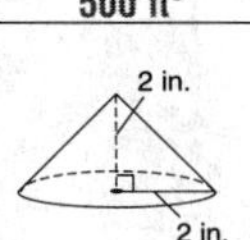
$B = 9.3$ cm²
6 cm³

6. 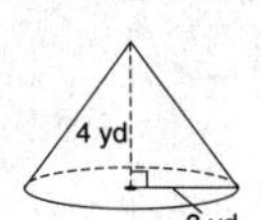
38 yd³

7. 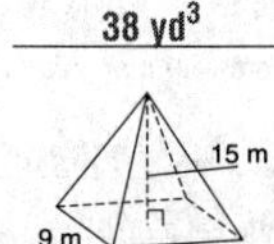
8 in³

8. 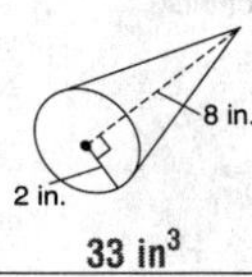
2,826 mm³

9. 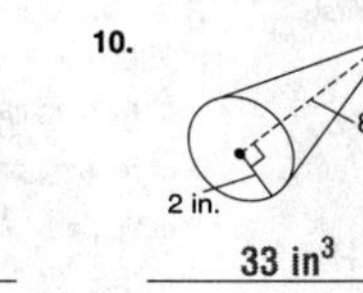
540 m³

10. 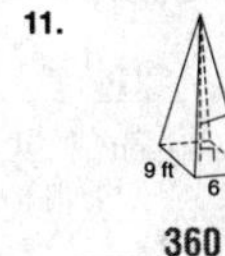
33 in³

11.
360 ft³

19
Holt Mathematics

Practice B
Volume of Pyramids and Cones

Find the volume of each pyramid to the nearest tenth.

1. 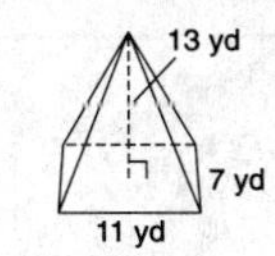
333.7 yd³

2. 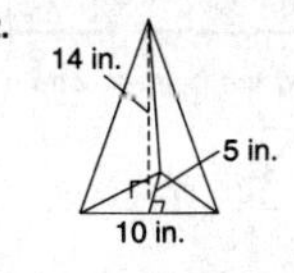
116.7 in³

3. 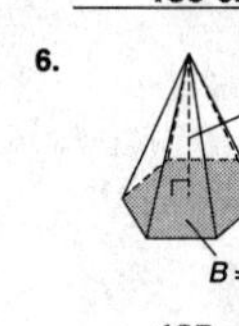
108 cm³

4. 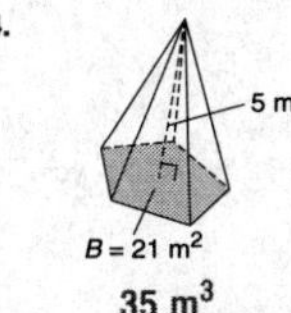
$B = 21$ m²
35 m³

5. 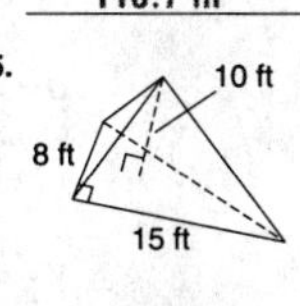
200 ft³

6.
$B = 35$ m²
105 m³

Find the volume of each cone to the nearest tenth.

7. 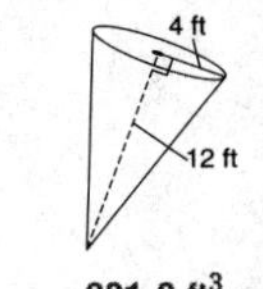
201.0 ft³

8. 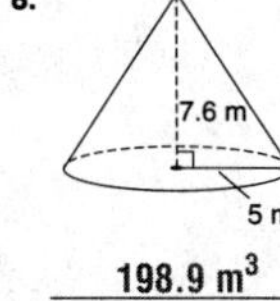
198.9 m³

9. 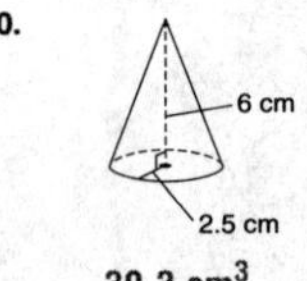
32.7 in³

10. 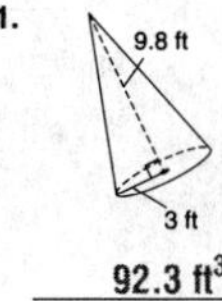
39.3 cm³

11.
92.3 ft³

12. 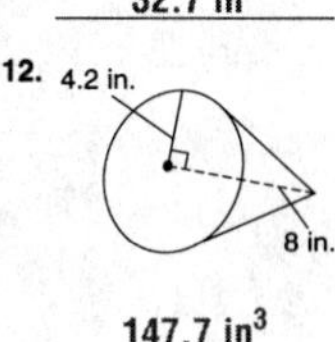
147.7 in³

20
Holt Mathematics

Practice C
Volume of Pyramids and Cones

Find the volume of each pyramid to the nearest tenth.

1. 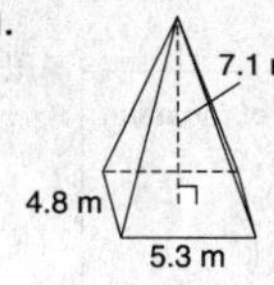
60.2 m³

2. 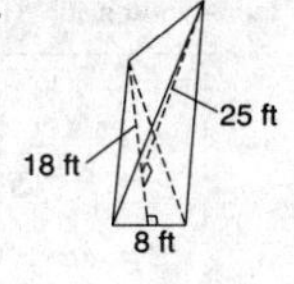
600 ft³

3. 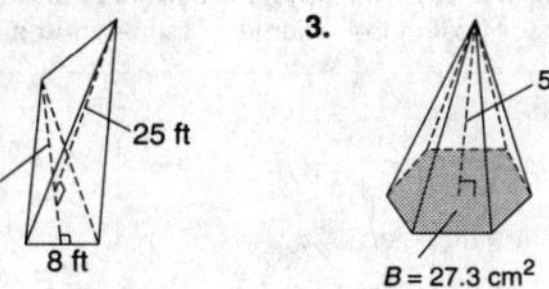
$B = 27.3$ cm²
49.1 cm³

Find the volume of each cone to the nearest tenth.

4. 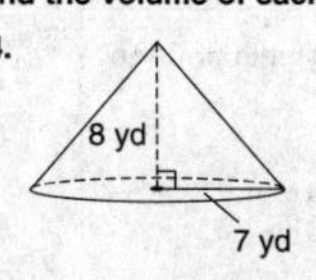
410.3 yd³

5. 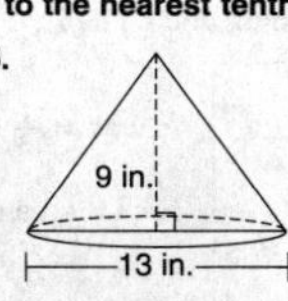
398.0 in³

6. 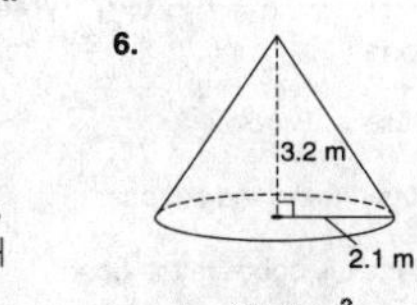
14.8 m³

Find the volume of each solid to the nearest tenth.

7. rectangular pyramid 7 ft by 3 ft by 9 ft high — 63 ft³

8. cone with radius 8 m and height 11 m — 736.9 m³

9. square pyramid with a 4 ft base and height 7 ft — 37.3 ft³

10. cone with diameter 15 in. and height 20 in. — 1,177.5 in³

11. cone with diameter 11 in. and height of 6 in. — 190 in³

12. triangular pyramid with height 10 cm and a base that is a right triangle with legs of 3 cm and 4 cm — 40 cm³

21
Holt Mathematics

Reteach
Volume of Pyramids and Cones

The volume of a prism with base area B and height h is $V = Bh$.

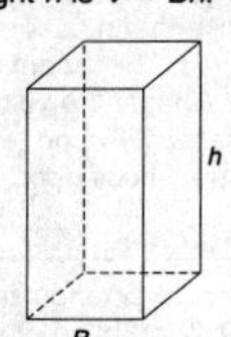

A pyramid with the same base B and height h has the volume $V = \frac{1}{3}Bh$.

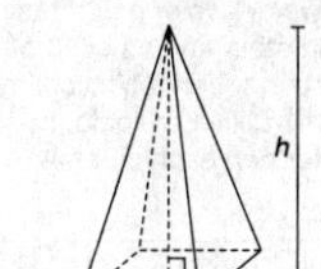

Find the volume of each pyramid.

1. The area of the base is
$B = \ell \cdot w = \underline{5} \cdot \underline{3} = \underline{15}$ ft²

2. The height of the pyramid is $\underline{6}$ ft.

3. $V = \frac{1}{3}Bh = \frac{1}{3} \cdot \underline{15} \cdot \underline{6} = \underline{30}$ ft³

4. 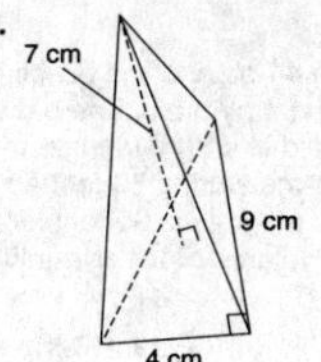
42 cm³

5. 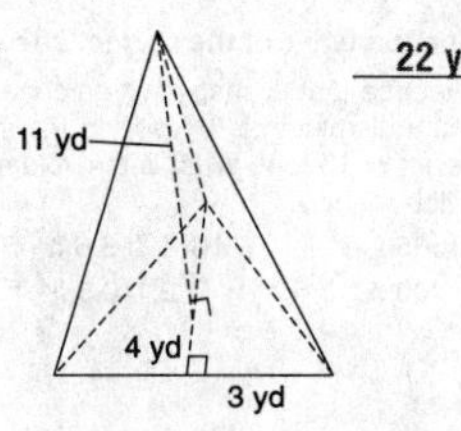
22 yd³

6. 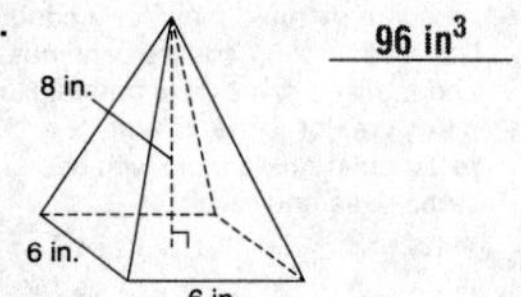
96 in³

7. 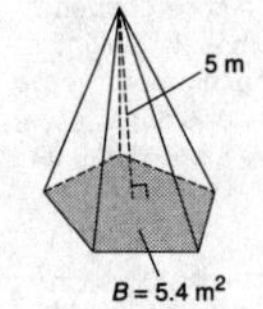
$B = 5.4$ m²
9 m³

22
Holt Mathematics

Reteach
Volume of Pyramids and Cones (continued)

You can find the volume of a cone by using the formula $V = \frac{1}{3}Bh$.

Find the volume of each cone to the nearest whole number.

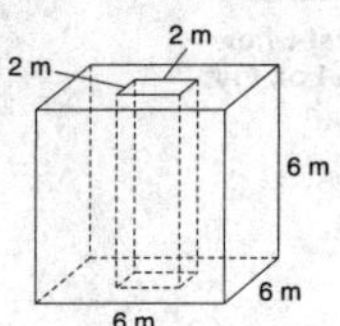

8. The area of the base is
$B = \pi r^2 = 3.14 \cdot \underline{\ 2\ }^2 = \underline{\ 13\ }$ m²

9. $V = \frac{1}{3}Bh = \frac{1}{3} \cdot \underline{\ 13\ } \cdot \underline{\ 5\ }$
$= \frac{1}{3} \cdot \underline{\ 65\ } = \underline{\ 22\ }$ m³

10. 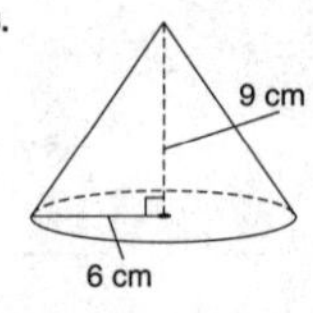

$B = \pi r^2$
$B = 3.14 \cdot \underline{\ 5\ }^2$
$\quad = 3.14 \cdot \underline{\ 25\ } = \underline{\ 79\ }$ yd²
$V = \frac{1}{3}Bh$
$V = \frac{1}{3} \cdot \underline{\ 79\ } \cdot \underline{\ 4\ }$
$\quad = \frac{1}{3} \cdot \underline{\ 316\ } = \underline{\ 105\ }$ yd³

11. 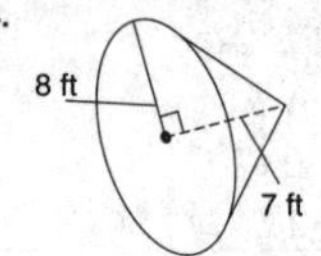

$\underline{75\ \text{cm}^3}$

12.

$\underline{733\ \text{ft}^3}$

13.

$\underline{339\ \text{cm}^3}$

14.

$\underline{469\ \text{ft}^3}$

23   **Holt Mathematics**

Challenge
Take It Out

Sometimes you can find the volume of an oddly shaped figure by first finding the volume of the entire solid and then subtracting a part that is missing.

Find the volume of each figure. Round to the nearest whole number.

1.

Cube with rectangular prism removed

V of cube: $\underline{\ 216\ \text{m}^3\ }$
V of missing part: $\underline{\ 24\ \text{m}^3\ }$
V of figure: $\underline{\ 192\ \text{m}^3\ }$

2.

Rectangular prism with pyramid removed

V of prism: $\underline{\ 640\ \text{ft}^3\ }$
V of missing part: $\underline{\ 213\ \text{ft}^3\ }$
V of figure: $\underline{\ 427\ \text{ft}^3\ }$

3.

Cube with 2 m × 2 m × 6 m corners removed

V of cube: $\underline{\ 216\ \text{m}^3\ }$
V of missing part: $\underline{\ 96\ \text{m}^3\ }$
V of figure: $\underline{\ 120\ \text{m}^3\ }$

4.

Rectangular prism with triangular prism removed

V of rectangular prism: $\underline{\ 960\ \text{cm}^3\ }$
V of missing part: $\underline{\ 240\ \text{cm}^3\ }$
V of figure: $\underline{\ 720\ \text{cm}^3\ }$

24   **Holt Mathematics**

Problem Solving
Volume of Pyramids and Cones

Write the correct answer.

1. Each of the cone-shaped cups near the water cooler has a radius of 3 centimeters and a height of 10 centimeters. If 1 cubic centimeter can hold 1 milliliter of liquid, how much water can each cup hold?

$\underline{94.2\ \text{mL}}$

2. The Great Pyramid in Egypt has a square base that measures 751 feet on each side. The pyramid is 481 feet high. What is the volume of the Great Pyramid? Round your answer to the nearest cubic foot.

$\underline{90{,}428{,}160\ \text{ft}^3}$

3. A waffle cone that holds ice cream is 15 centimeters high and has a diameter of 10 centimeters. What volume of ice cream can it hold if it is filled to the top?

$\underline{392.5\ \text{cm}^3}$

4. The base of a rectangular prism is congruent to the base of a pyramid. The height of the pyramid is 3 times the height of the prism. Which figure has a greater volume? Explain.

$\underline{\text{The figures have the same}}$
$\underline{\text{volume. } V(\text{prism}) = Bh;}$
$\underline{V(\text{pyramid}) = (\frac{1}{3}Bh) \cdot 3 = Bh}$

Choose the letter of the correct answer.

5. A teepee that is shaped like a cone has a diameter of 12 feet and a height of 15 feet. What is the volume of the teepee?
Ⓐ 565.2 ft³   C 1,695.6 ft³
B 706.5 ft³   D 2,119.5 ft³

6. The top of a 44-story office building is shaped like a pyramid. The base of the pyramid is a right triangle with the two legs measuring 73 feet and 78 feet. The pyramid is 35 feet high. What is the volume of the pyramid?
F 199,290 ft³   H 41,756 ft³
G 66,430 ft³   Ⓙ 33,215 ft³

7. The diameter of a cone-shaped container is 4 inches. Its height is 6 inches. How much greater is the volume of a cylinder-shaped container with the same diameter and height?
Ⓐ 50.24 in³   C 100.98 in³
B 75.36 in³   D 200.96 in³

8. A square pyramid mold for a candle has a base of 64 square centimeters and a height of 12 centimeters. How much greater is the volume of a rectangular prism mold with the same base and height?
Ⓕ 64 cm³   H 256 cm³
G 96 cm³   Ⓙ 512 cm³

25   **Holt Mathematics**

Reading Strategies
Use a Graphic Organizer

This chart will help you compare pyramids and cones and the formulas for finding the volume of these figures.

Pyramid	Volume of Pyramid
• Base is a polygon.	Volume = $\frac{1}{3}B \cdot h$
• Faces are triangles.	
• Vertex and base are at opposite ends.	(B = area of base)

Pyramids and Cones

Cone	Volume of Cone
• Base is a circle.	Volume = $\frac{1}{3}\pi r^2 \cdot h$
• Has one curved surface.	(πr^2 = area of base)
• Vertex is opposite the base.	

Use the graphic organizer to answer each question.

1. Which figure has one curved surface? $\underline{\text{a cone}}$

2. What forms the base of a pyramid? $\underline{\text{a polygon}}$

3. What forms the base of the cone? $\underline{\text{a circle}}$

4. What is the formula for the area of a cone's base?

$\underline{\pi r^2}$

5. What shape are the faces of a pyramid?

$\underline{\text{triangle}}$

6. Compare the volume formulas. How are they alike?

$\underline{\text{In each you find } \frac{1}{3} \text{ of the area of the base multiplied by the height.}}$

26   **Holt Mathematics**

Puzzles, Twisters & Teasers
Loud and Clear!

Circle words from the list in the word search. Then find a word that answers the riddle. Circle it and write it on the line.

volume   pyramid   cone   sphere   container
prism   congruent base   radius   cylinder

How does the sky listen to music?

Through a ___cloud___ -speaker

27
Holt Mathematics

Practice A
Surface Area of Prisms and Cylinders

Find the surface area of the prism formed by each net. Choose the letter for the best answer.

1.
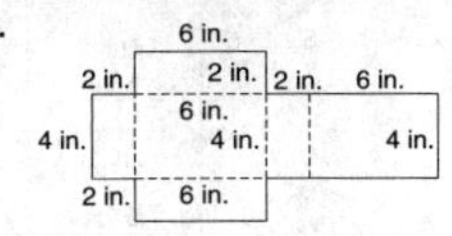

2.

A 36 in²   C 48 in²
B 44 in²   (D) 88 in²

F 95 cm²   (H) 190 cm²
G 100 cm²   J 380 cm²

Find the surface area of the cylinder formed by each net to the nearest whole number. Choose the letter for the best answer.

3.

4.
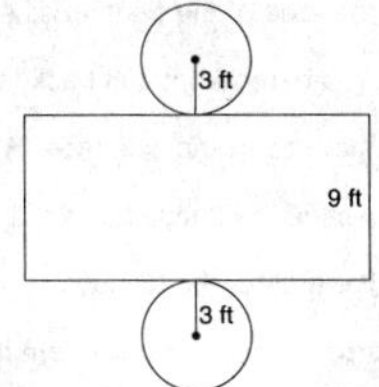

A 157 cm²   (C) 471 cm²
B 314 cm²   D 942 cm²

(F) 226 ft²   H 113 ft²
G 170 ft²   J 57 ft²

5. A soup can has a diameter of 8 centimeters and a height of 10 centimeters. The label for the can covers the entire side of the can. About how many square centimeters of paper are needed for the label? Round to the nearest whole number.

251 cm²

28
Holt Mathematics

Practice B
Surface Area of Prisms and Cylinders

Find the surface area of the prism formed by each net to the nearest tenth.

1.

2.

142 in²

600 mm²

Find the surface area of the cylinder formed by each net to the nearest tenth.

3.
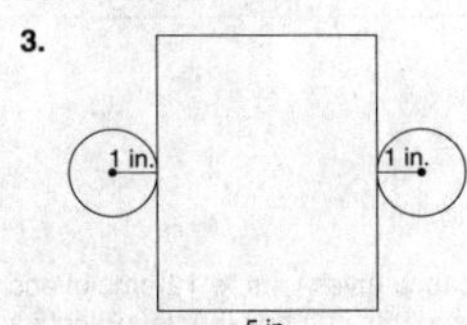

4.
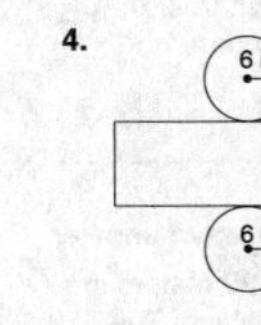

37.7 in²

678.2 in²

5. Mr. Wang has a circular swimming pool with a diameter of 15 feet and a height of 5 feet. Mr. Wang buys a liner to cover the bottom and the sides of the pool. To the nearest square foot, about how many square feet of liner should Mr. Wang buy in order to have enough liner? Explain your answer.

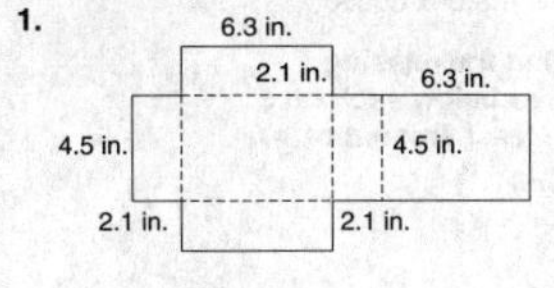

He needs 412.125 ft². So, to the nearest foot, he should buy 413 ft².

29
Holt Mathematics

Practice C
Surface Area of Prisms and Cylinders

Find the surface area of the prism formed by each net to the nearest tenth.

1.

2.

102.1 in²

217.7 cm²

Find the surface area of the cylinder formed by each net to the nearest tenth.

3.

4.
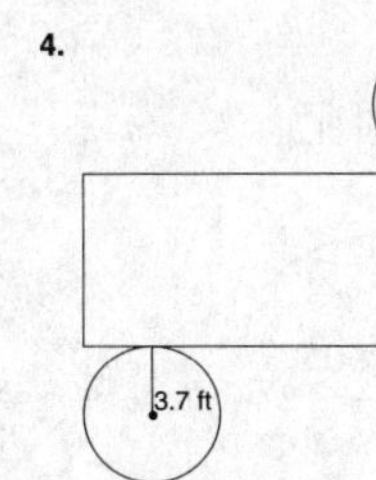

525.3 cm²

302.1 ft²

5. The diagram shows a closet that is shaped like a rectangular prism. Judith is lining the inside of the closet, including the floor, with cedar. Judith is not lining the doorway. If cedar costs $4.00 per square foot, what will the total cost of the cedar be?

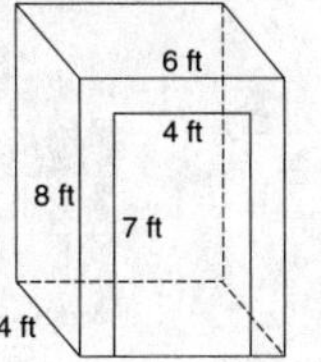

$720

30
Holt Mathematics

Reteach
Surface Area of Prisms and Cylinders

The surface area of a three-dimensional figure is the combined areas of the faces. You can find the surface area of a prism by drawing a **net** of the flattened figure.

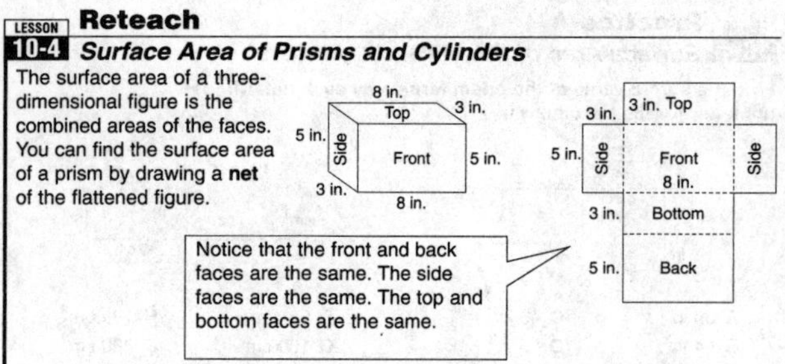

Notice that the front and back faces are the same. The side faces are the same. The top and bottom faces are the same.

Find the surface area of the prism formed by each net.

1. Find the area of the front face: $A = \underline{8} \cdot \underline{5} = \underline{40}$ in^2.

   The area of the front and back faces is $2 \cdot \underline{40} = \underline{80}$ in^2.

2. Find the area of the side face: $A = \underline{5} \cdot \underline{3} = \underline{15}$ in^2.

   The area of the 2 side faces is $2 \cdot \underline{15} = \underline{30}$ in^2.

3. Find the area of the top face: $A = \underline{8} \cdot \underline{3} = \underline{24}$ in^2.

   The area of the top and bottom faces is $2 \cdot \underline{24} = \underline{48}$ in^2.

4. Combine the areas of the faces: $\underline{80} + \underline{30} + \underline{48} = \underline{158}$ in^2.

   The surface area of the prism is $\underline{158}$ in^2.

Find the surface area of the prism formed by each net.

5.
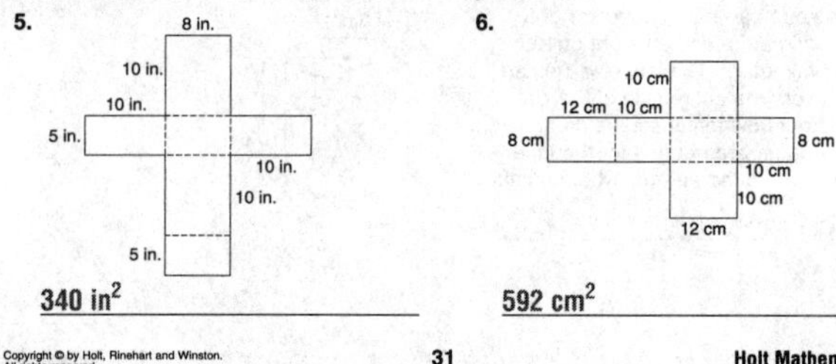

340 in^2

6.

592 cm^2

Holt Mathematics

Reteach
Surface Area of Prisms and Cylinders (continued)

The surface area of a cylinder consists of a rectangle and 2 circular bases. The width of the rectangle is equal to the height of the cylinder. The length of the rectangle is equal to the circumference of the bases.

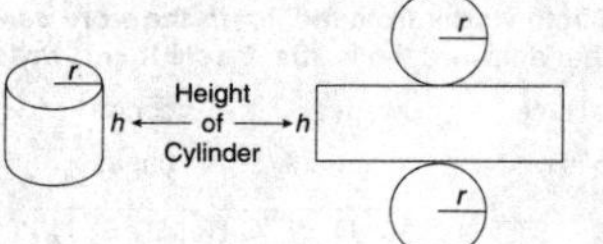

Find the surface area of the cylinder formed by the net to the nearest whole number.

7. Find the area of each circular base.

   $A = \pi r^2 = 3.14 \cdot \underline{5^2} = \underline{79}$ cm^2

8. Find the area of both circular bases.

   $A = 2 \cdot \underline{79} = \underline{158}$ cm^2

9. Find the area of the rectangle.

   $A = \ell \cdot w$

   $A = 2\pi r \cdot h$

   $A = 2 \cdot 3.14 \cdot \underline{5} \cdot \underline{8}$

   $A = \underline{251}$ cm^2

10. Combine the areas of the circular bases and the area of the rectangle to find the surface area, S, of the cylinder.

    $S = \underline{158} + \underline{251} = \underline{409}$ cm^2

Find the surface area of the cylinder formed by each net. Round to the nearest whole number. Use 3.14 for π.

11.
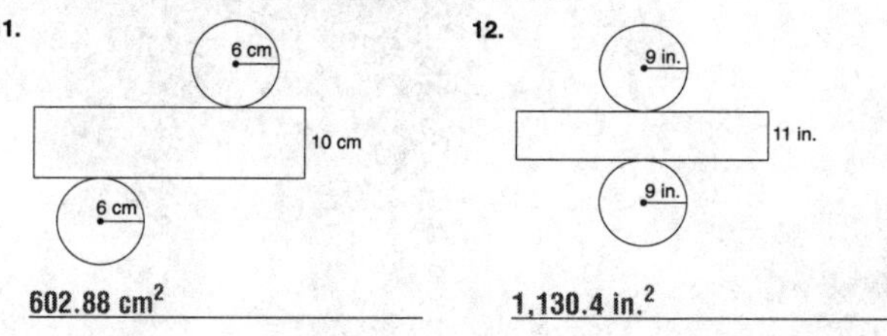

602.88 cm^2

12.

1,130.4 in.^{2}

Holt Mathematics

Challenge
Saving Trees

Boxes with different dimensions can have the same volume. Marcus Manufacturing uses boxes to ship office supplies to stores. The less cardboard used to make a box, the less the box costs.

In each set, the boxes have equal volume. Find the missing dimension, if necessary. Write the surface area below each box. Then circle the box in each set that uses the least amount of cardboard.

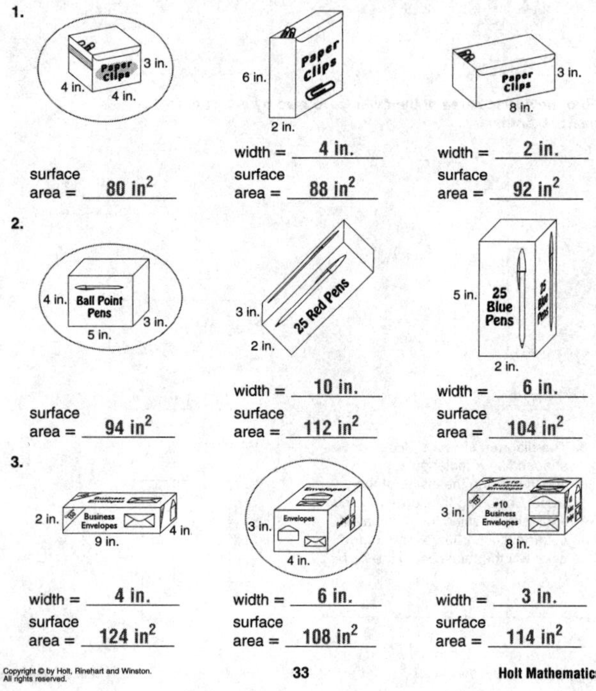

1.

surface area = $\underline{80}$ in^2

width = $\underline{4}$ in.
surface area = $\underline{88}$ in^2

width = $\underline{2}$ in.
surface area = $\underline{92}$ in^2

2.

surface area = $\underline{94}$ in^2

width = $\underline{10}$ in.
surface area = $\underline{112}$ in^2

width = $\underline{6}$ in.
surface area = $\underline{104}$ in^2

3.

width = $\underline{4}$ in.
surface area = $\underline{124}$ in^2

width = $\underline{6}$ in.
surface area = $\underline{108}$ in^2

width = $\underline{3}$ in.
surface area = $\underline{114}$ in^2

Holt Mathematics

Problem Solving
Surface Area of Prisms and Cylinders

Write the correct answer.

1. A can of peas is 3 inches in diameter and 4.5 inches tall. What is the area of the label used around the can?

   42.39 in^2

2. How much wrapping paper do you need to completely cover a rectangular box that is 20 inches by 18 inches by 2 inches?

   872 in^2

3. Jan puts frosting on a circular cake. The cake has three layers, each with a diameter of 20 centimeters and a height of 5 centimeters. Jan puts frosting between the layers and on the outside, except for the bottom. What is the area that Jan frosts? Round to the nearest square centimeter.

   1,884 cm^2

4. A cardboard storage carton has a length of 3 feet, a width of 2 feet, and a volume of 12 ft^3. What is the minimum amount of cardboard needed to make the box?

   32 ft^2

Choose the letter of the correct answer.

5. A cylindrical building is 30 meters in diameter and 50 meters high. The outside of the building, excluding the roof, is completely covered in glass. To the nearest square foot, what is the total area of the glass?

   A 1,413 m^3 C 6,113 m^3
   B 4,710 m^3 D 9,420 m^3

6. Rebecca gives gifts to 12 employees. Each gift is in a box that is 12 inches by 10 inches by 3 inches. How much wrapping paper does Rebecca need to completely the cover the boxes?

   F 360 in^2 H 4,320 in^2
   G 1,720 in^2 J 14,400 in^2

7. A cylinder-shaped sculpture is 24 meters high with a diameter of 6.8 meters. An artist plans to spray-paint the entire surface with silver paint. If one can of spray paint covers 50 square meters, how many cans does the artist need to paint the sculpture?

   A 51 cans   C 12 cans
   B 22 cans   D 10 cans

8. A rectangular sofa cushion is 36 inches by 30 inches by 20 inches. How many cushions can be covered with 38,400 square inches of material?

   F 6 cushions   H 9 cushions
   G 8 cushions   J 10 cushions

Holt Mathematics

Holt Mathematics

Reading Strategies
Analyze Information

If you unfold a three-dimensional figure and lay it flat, you have
made a net.

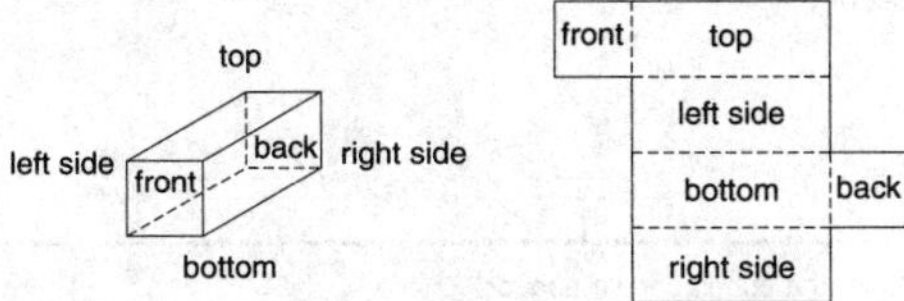

A **net** is a two-dimensional shape that lets you picture all the faces
of a three-dimensional figure. A net helps you see how much
surface a three-dimensional figure covers.

The **surface area** of a three-dimensional figure is the total area of
all its faces.

If you analyze the net of a rectangular prism, you notice there are
six faces. Each face pairs up with another, congruent face:

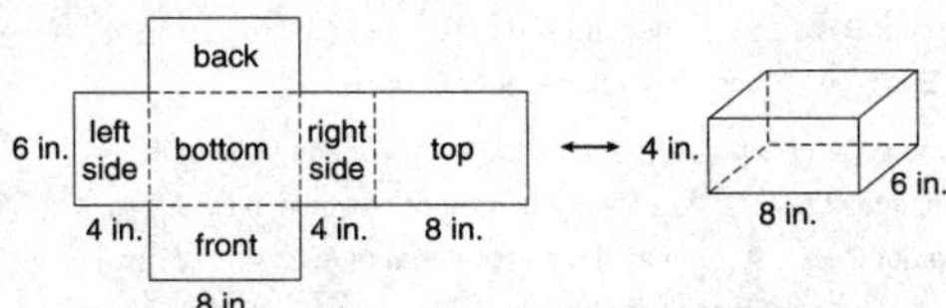

To find the surface area of a rectangular prism, find the sum of the
areas of the six faces (rectangles). *Hint:* $A = \ell w$.

Answer the following to find the surface area of the rectangular prism.

1. What is the area of the bottom rectangle? _____ 48 in^2
2. What is the area of the top rectangle? _____ 48 in^2
3. What is the area of the back rectangle? _____ 32 in^2
4. What is the area of the front rectangle? _____ 32 in^2
5. What is the area of the right rectangle? _____ 24 in^2
6. What is the area of the left rectangle? _____ 24 in^2
7. What is the surface area of the rectangular prism? _____ 208 in^2

Holt Mathematics

Puzzles, Twisters & Teasers
Bee a Math Marvel!

Find the surface area of each figure below.
Round to the nearest tenth. Use your answers
to solve the riddle.

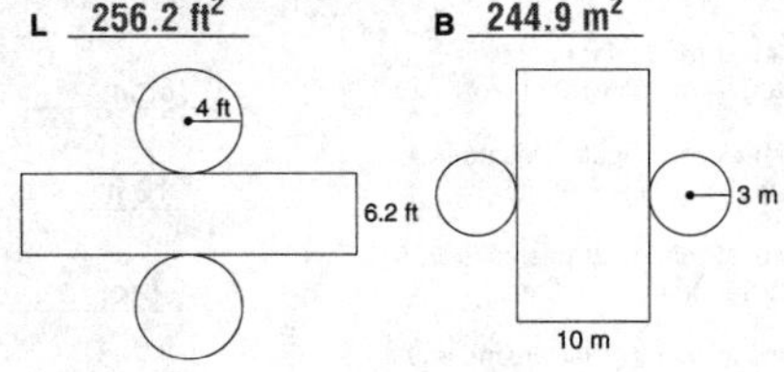

L _256.2 ft^2_   **B** _244.9 m^2_

A _562 m^2_   **H** _653.1 in^2_   **T** _286 ft^2_

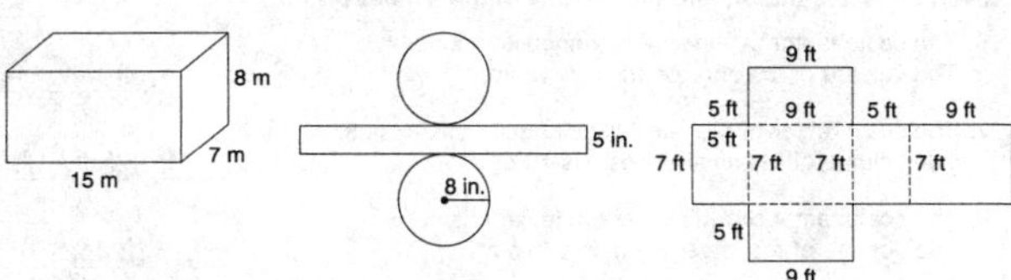

O _351.7 ft^2_   **S** _1,160 ft^2_   **I** _264 m^2_

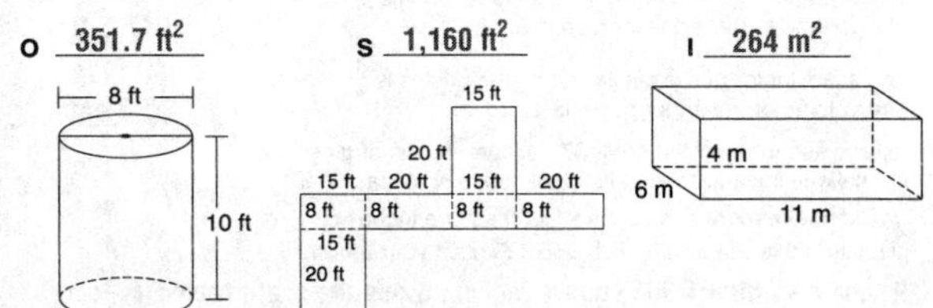

Why couldn't the bee go to the dance?

I T   W A S   A   M O T H B A L L
264 286   562 1,160 562   351.7 286 653.1 244.9 562 256.2 256.2

Holt Mathematics

Practice A
Changing Dimensions

Two prisms are similar. The scale factor and the surface area of
the smaller prism are given in Column A. Draw a line to the
surface area of the larger prism in Column B.

Column A Prisms

1. scale factor: 3
   surface area of the
   smaller prism: 9 cm^2

2. scale factor: 4
   surface area of the
   smaller prism: 5 cm^2

3. scale factor: 5
   surface area of the
   smaller prism: 4 cm^2

4. scale factor: 2
   surface area of the
   smaller prism: 24 cm^2

Column B Prisms

A 96 cm^2
B 90 cm^2
C 80 cm^2
D 100 cm^2
E 81 cm^2

Two prisms are similar. The scale factor and the volume of the
smaller prism are given in Column A. Draw a line to the volume
of the larger prism in Column B.

Column A Prisms

5. scale factor: 3
   volume of the smaller
   prism: 5 ft^3

6. scale factor: 4
   volume of the smaller
   prism: 2 ft^3

7. scale factor: 5
   volume of the smaller
   prism: 1 ft^3

8. scale factor: 2
   volume of the smaller
   prism: 17 ft^3

Column B Prisms

F 125 ft^3
G 128 ft^3
H 135 ft^3
I 136 ft^3
J 144 ft^3

9. A fish tank holds 6 gallons of water. A larger, similarly
   shaped fish tank has a scale factor of 2. How many
   times more water does the larger fish tank hold than
   the smaller fish tank?   _8 times more_

Holt Mathematics

Practice B
Changing Dimensions

Given the scale factor, find the surface area of the similar
prism.

1. The scale factor of two similar rectangular prisms is 3.
   The surface area of the smaller prism is 15 in^2.   _135 in^2_

2. The scale factor of two similar triangular prisms is 2.
   The surface area of the smaller prism is 25 cm^2.   _100 cm^2_

3. The scale factor of two similar rectangular prisms is $\frac{1}{2}$.
   The surface area of the larger prism is 960 ft^2.   _240 ft^2_

4. The scale factor of two similar triangular prisms is 5.
   The surface area of the smaller prism is 10 m^2.   _250 m^2_

5. The scale factor of two similar rectangular prisms is $\frac{1}{5}$.
   The surface area of the larger prism is 625 in^2.   _25 in^2_

6. The scale factor of two similar pentagonal prisms is 4.
   The surface area of the smaller prism is 16 cm^2.   _256 cm^2_

Given the scale factor, find the volume of the similar prism.

7. The scale factor of two similar triangular prisms is 3.
   The volume of the smaller prism is 8 in^3.   _216 in^3_

8. The scale factor of two similar rectangular prisms is $\frac{1}{2}$.
   The volume of the larger prism is 648 m^3.   _81 m^3_

9. The scale factor of two similar triangular prisms is 4.
   The volume of the smaller prism is 10 cm^3.   _640 cm^3_

10. The scale factor of two similar rectangular prisms is $\frac{1}{4}$.
    The volume of the larger prism is 1,920 ft^3.   _30 ft^3_

11. The scale factor of two similar triangular prisms is 2.
    The volume of the smaller prism is 72 yd^3.   _576 yd^3_

12. A small tank weighs 24 pounds when it is full of water.
    A larger tank that is similar in shape has a scale factor
    of 3. How much does the larger tank weigh when filled
    with water?   _648 lb_

Holt Mathematics

Practice C
Changing Dimensions

Given the scale factor, find the surface area of the similar prism.

1. The scale factor of two similar rectangular prisms is 5. The surface area of the smaller prism is 13 in^2.

   325 in^2

2. The scale factor of two similar triangular prisms is $\frac{1}{8}$. The surface area of the larger prism is 1,024 cm^2.

   16 cm^2

3. The scale factor of two similar rectangular prisms is 4. The surface area of the smaller prism is 32 ft^2.

   512 ft^2

4. The scale factor of two similar triangular prisms is $\frac{1}{3}$. The surface area of the larger prism is 828 m^2.

   92 m^2

5. The scale factor of two similar rectangular prisms is 3. The surface area of the smaller prism is 13.5 yd^2.

   121.5 yd^2

Given the scale factor, find the volume of the similar prism.

6. The scale factor of two similar triangular prisms is 5. The volume of the smaller prism is 12 in^3.

   $1,500$ in^3

7. The scale factor of two similar rectangular prisms is 3. The volume of the smaller prism is 75 m^3.

   $2,025$ m^3

8. The scale factor of two similar triangular prisms is $\frac{1}{2}$. The volume of the larger prism is 976 cm^3.

   122 cm^3

9. The scale factor of two similar rectangular prisms is 5. The volume of the smaller prism is 2.5 ft^3.

   312.5 ft^3

10. The scale factor of two similar triangular prisms is $\frac{1}{4}$. The volume of the larger prism is 1,120 cm^3.

    17.5 cm^3

11. A large fish tank is made of 2,376 square inches of glass. A smaller, similarly shaped fish tank has a scale factor of $\frac{1}{2}$. Did it take more or less than 1,000 square inches of glass to make the smaller fish tank? Explain your answer.

    It took less than 1,000 square inches. Since the scale factor is $\frac{1}{2}$, the

    surface area of the smaller tank is $\frac{1}{2} \cdot \frac{1}{2}$, or $\frac{1}{4}$ as great as 2,376.

    $\frac{1}{4} \cdot 2.376 = 594$, and $594 < 1,000$.

39

Holt Mathematics

Reteach
Changing Dimensions

Changing the dimensions of a three-dimensional figure affects its surface area and its volume.

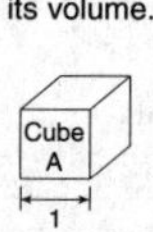
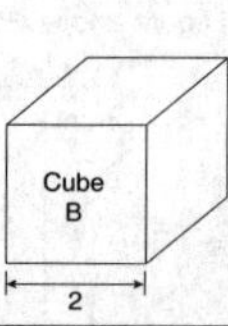
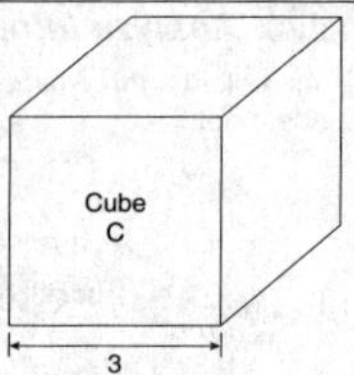

1. Complete the table for the cubes shown above.

Length of Side	Area of One Face	Surface Area	Volume
1	1 unit^2	6 units^2	1 unit^3
2	4 units^2	24 units^2	8 units^3
3	9 units^2	54 units^2	27 units^3

2. The length of one edge of B is __2__ times the length of one edge of A.

   The area of one face of B is __4__ times the area of one face of A.

   The surface area of B is __4__ times the surface area of A: $2^2 = $ __4__.

   The volume of B is __8__ times the volume of A: $2^3 = $ __8__.

3. The length of one edge of C is __3__ times the length of one edge of A.

   The area of one face of C is __9__ times the area of one face of A.

   The surface area of C is __9__ times the surface area of A: $3^2 = $ __9__.

   The volume of C is __27__ times the volume of A: $3^3 = $ __27__.

For any set of similar three-dimensional figures:

- The surface area of the larger figure equals the surface area of the smaller figure times the scale factor squared.
- The volume of the larger figure equals the volume of the smaller figure times the scale factor cubed.

The scale factor of two similar prisms is 4.

4. The surface area of the smaller prism is 9 in^2. Multiply by __4^2__, or __16__.

   The surface area of the larger prism is __144__ in^2.

40

Holt Mathematics

Challenge
Eggs-actly

The volumes of two similar-shaped figures are proportional to the cube of their lengths.

Suppose an extra-large egg is 2.5 inches long, compared to a smaller egg 1.25 inches long. What is the ratio of their volumes?

Cube the ratio of their lengths.

$$\left(\frac{2.5}{1.25}\right)^3 = \left(\frac{2}{1}\right)^3 = \frac{8}{1}$$

The ratio of their volumes is 8:1.

So, the larger egg has 8 times as much volume, 8 times as much weight, and 8 times as much food.

Solve. Show your work.

1. One egg is 2 inches long. Another is 2.75 inches long. How many times greater is the volume of the larger egg?

   $$\left(\frac{2.75}{2}\right)^3 \approx \frac{20.8}{8} \approx 2.6; \text{ about 2.6 times}$$

2. One egg is 2.25 inches long. Another is 1.5 inches long. How many times greater is the weight of the larger egg?

   $$\left(\frac{2.25}{1.5}\right)^3 \approx \left(\frac{3}{2}\right)^3 \text{ about 3.4 times}$$

3. The ratio of the lengths of two eggs is 5 to 2. The larger egg contains how many times the food of the smaller egg?

   $$\left(\frac{5}{2}\right)^3 = \frac{125}{8} = 15.625; \text{ about 15.6 times as much}$$

4. An egg is 2 inches long. What is the approximate length of an egg with twice the volume?

   about 2.5 in.

5. An egg is 3 inches long. What is the approximate length of an egg with half the weight?

   about 2.4 in.

6. An egg is 1.75 inches long. What is the approximate length of an egg with twice the food?

   about 2.2 in.

41

Holt Mathematics

Problem Solving
Changing Dimensions

Write the correct answer.

1. In the late 1800s, wax cylinders were used to record sound. The surface area of the curved side of each cylinder was about 25 square inches. If a larger model of this cylinder is created using a scale factor of 3, what is the lateral surface area of the model?

   about 225 in^2

2. A 5-foot wide, 100-foot tall cylindrical water tower was built in St. Louis in the early 1800s. An architecture student wants to build a model using a scale factor of $\frac{1}{6}$. What will be the volume of the model to the nearest cubic foot?

   9 ft^3

3. A cone-shaped plastic cup holds 24 ounces of water. A smaller cup has a scale factor of $\frac{1}{2}$. How much water does the smaller cup hold?

   3 oz

4. In the game of Ring Taw, players use a shooting marble that has a surface area of 1.77 square inches. What is the surface area of a large ball if the scale factor is 5?

   44.25 in^2

Choose the letter of the correct answer.

5. The volume of a rectangular prism is 48 cubic centimeters. The volume of a similar rectangular prism is 6 cubic centimeters What is the scale factor for the rectangles?

   A 8

   B 6

   C 4

   (D) 2

6. A cooking pot used in the cafeteria weighs 64 pounds when it is filled with soup. How much would a similar pot with a scale factor of $\frac{1}{2}$ weigh when filled with the same soup?

   F 4 lb

   (G) 8 lb

   H 16 lb

   J 32 lb

7. For his science project, Marty is building a model of Pluto, which has a surface area of about 6,376,000 square miles. He plans to cover his model with red foil. If he uses a scale factor of 5,000, how much red foil will he need to the nearest hundredth of a square mile?

   A 1.26 mi^2

   B 0.026 mi^2

   (C) 0.26 mi^2

   D 0.126 mi^2

8. The scale factor of two similar triangular prisms is 5. What is the possible surface area of both prisms?

   (F) 2,000 cm^2 and 80 cm^2

   G 125 cm^2 and 25 cm^2

   H 5,000 cm^2 and 40 cm^2

   J 10,000 cm^2 and 60 cm^2

42

Holt Mathematics

Reading Strategies
Organize Information

Each face on this cube measures 1 unit by 1 unit. The area of one face is 1 square unit.

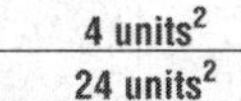

Adding 6 faces = 6 square units. The surface area is 6 units^2.

Each face on this cube measures 2 units by 2 units.

1. What is the area of one face? — 4 units^2

2. What is the total area of all six faces? — 24 units^2

You can use a table to organize the information about the two cubes.

Dimensions of Each Face	Area of 1 Face	Total Surface Area
1 unit by 1 unit	1 unit^2	6 • 1 = 6 units^2
2 units by 2 units	4 units^2	6 • 4 = 24 units^2

This cube would be the next cube in the list.

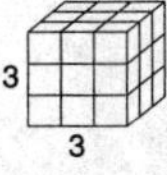

Use this cube to answer the following questions.

3. What are the dimensions of each face on the cube? — 3 units by 3 units

4. What is the area of one face of the cube?

   3 units • 3 units = 9 square units or 9 units^2

5. How many faces does the cube have? — 6

6. What is the surface area for the cube?

   6 units • 9 units = 54 units^2

7. If the list above continued, what would be the dimensions of each face on the next cube? — 4 units by 4 units

43

Holt Mathematics

Puzzles, Twisters & Teasers
A New Dimension!

Complete both charts. Use your answers to solve the riddle.

scale factor	smaller surface area	larger surface area	
4	35 in^2	560 in^2	T
6	24 m^2	864 m^2	S
8	40 ft^2	2,560 ft^2	U
9	20 m^2 H	1.602 m^2	

scale factor	smaller volume	larger volume	
2	29 ft^3	232 ft^3	C
3	18 lb^3	486 lb^3	E
5	8 cm^3	1,000 cm^3	R
4	17 in^3	1,088 in^3	O
4	32 lb^3	2,048 lb^3	A
2	75 in^3 K	600 in^3	

Why did the hairdresser win the race?

S H E   T O O K   A
864 20 486   560 1,088 1,088 75   2,048

S H O R T C U T.
864 20 1,088 1,000 560   232 2,560 560

44

Holt Mathematics